# Optimization in industry

# Optimization in industry

## VOLUME ONE
## Optimization techniques

# T. A. J. Nicholson

Routledge
Taylor & Francis Group

LONDON AND NEW YORK

First published 2007 by Transaction Publishers

Published 2017 by Routledge
2 Park Square, Milton Park, Abingdon, Oxon OX14 4RN
711 Third Avenue, New York, NY 10017, USA

*Routledge is an imprint of the Taylor & Francis Group, an informa business*

Library of Congress Catalog Number: 2007019719

Library of Congress Cataloging-in-Publication Data

Nicholson, T.A.J.
 Optimization in industry / T.A.J. Nicholson.
  p. cm.
 Originally published: Chicago : Aldine Pub. Co., 1974.
 Includes index.
 ISBN 978-0-202-30965-1 (acid-free paper)
  1. Programming (Mathematics)  2. Mathematical optimization. I. Title.

T57.7.N53  2007
658.4'033—dc22—dc22                                        2007019719


ISBN 13: 978-0-202-30965-1 (pbk)

*For Claire
Graham, and Diana,
and all other
optimists*

# Acknowledgements

I am grateful to Professor Moore of the London Business School for his frequent encouragement in the course of writing this book. The radical changes and improvements in the text since its first draft are largely due to the students of the first two M.Sc. courses held at the Business School and members of staff, particularly to Stewart Hodges: I am extremely grateful to them and to all other critics although my appreciation was rather less at the time of the suggested alterations. My thanks are very much due to my mother for meticulously inserting all the corrections in an already messy typescript. I could not face the proof reading alone, and I must thank Roger Garside, formerly of the A.E.R.E. Harwell, for undertaking the bulk of the work at that stage.

# Contents

Foreword     xi

Preface     xiii

**1 The study of optimization**     1
    1.1   Mathematics and practitioners     1
    1.2   Applying optimization methods     3

**2 Mathematical background**     5
    2.1   Variables and functions     5
    2.2   Differentiation and integration of functions     6
    2.3   Vectors     8
    2.4   Definition of regions     10
    2.5   Matrices     11
    2.6   Permutations     12
    2.7   Set notation     12
    2.8   Definition of a network     13
    2.9   Probability distributions     14

**3 Optimization problems and methods**     17
    3.1   Basic components     17
    3.2   Problem variables     17
    3.3   Objective functions     18
    3.4   Constraints     19
    3.5   Equality constraints     19
    3.6   Inequality constraints     20
    3.7   Discrete constraints     21

3·8    Disjoint alternative regions                                         22
3.9    The general optimization problem                              23
3.10   Global optima                                                          23
3.11   Local optima                                                            23
3.12   The search for the optimum                                      24
3.13   Optimization methods                                               25

**4 Calculus and Lagrange multipliers**                               29
4.1    The achievement of the calculus                                 29
4.2    Restrictions on the use of the calculus                        30
4.3    Problems in one variable                                          31
4.4    Unconstrained problems in two or more variables        33
4.5    Equality constraints: Lagrange multipliers                  35
4.6    Sufficient conditions for local maxima and minima      37
4.7    Proof of the Lagrange multiplier method                    39

**5 Linear programming**                                                    42
5.1    Linear functions                                                       42
5.2    The standard form                                                    43
5.3    Converting to the standard form                                43
5.4    The ideas of the simplex method                              45
5.5    The canonical form                                                   47
5.6    Improving the basis                                                  49
5.7    Transformation to the new canonical form               51
5.8    Summary of the procedure for a change of basis        51
5.9    Obtaining an initial solution                                     54
5.10   The simplex tableau                                                 56
5.11   The revised simplex method                                     58
5.12   Difficulties in special cases                                      61
5.13   The fundamental theorem of linear programming       61
5.14   The convex space of feasible solutions                      63

**6 Optimization of non-linear functions**                            65
6.1    Non-linear optimization techniques                           65
6.2    Local and global optima                                          66
6.3    Finding an initial feasible solution                            67
6.4    Unconstrained problems: the failure of the calculus    69
6.5    Steepest descent for unconstrained problems            69
6.6    More advanced descent methods for unconstrained problems   72
6.7    Projected gradient methods for linear constraints      75
6.8    Created response surface techniques for general constraints   80
6.9    Direct search procedures                                         84
6.10   Separable programming                                           88
6.11   Approximation programming                                    91

6.12 The direction of steepest descent 95
6.13 Construction of the projection matrix 96
6.14 Convergence of the response surface technique 97

**7 Dynamic programming** 101
7.1 The aim of dynamic programming 101
7.2 Multi-stage decision processes 102
7.3 The dynamic programming method: forwards calculation 103
7.4 The backwards calculation 105
7.5 Network problems 106
7.6 Allocation of a single resource to a number of activities 108
7.7 Reliability problems 113
7.8 Computational comments 114
7.9 Allocation processes involving two types of resources or two
    constraints 116
7.10 Problems with two neighbouring stages related:
    backward calculations 118
7.11 A stochastic problem 119

**8 Branch and bound methods** 123
8.1 Solutions as tree searches 123
8.2 Rules for the branch and bound method 125
8.3 The three-machine scheduling problem 127
8.4 The knapsack problem 131
8.5 Integer programming 134
8.6 Computational considerations 138
8.7 Suboptimization 138

**9 Permutation procedures** 141
9.1 The need to suboptimize 141
9.2 Permutation problems 141
9.3 Locally optimal permutations 143
9.4 Calculation of locally optimal permutations 145
9.5 Choosing sets of exchanges 149
9.6 Multi-permutation problems 149

**10 Heuristic techniques** 152
10.1 The need for skilful guessing 152
10.2 The nature of allocation problems 153

**11 Problem specification and mathematical treatment** 161
11.1 Flexibility in problem formulation 161
11.2 The choice of the problem variables 162
11.3 Transferring factors between the objective and constraint
    functions 162

11.4 Introducing auxiliary variables                    164
11.5 Transformations                                     166
11.6 Duality in linear problems                          168
11.7 Interpretation of the dual problem                  172
11.8 Proof of the duality theorem                        174

Answers to exercises                                     178

# Foreword

This series arises directly out of the experience of teachers concerned with a range of (post-)graduate and post-experience courses, from Masters and Ph.D programmes, taught primarily to recent graduates of extremely high calibre but from very diverse disciplines, to programmes for executives, often graduates of some years' standing, in mid-career in management. In developing all of our courses from scratch (the school opened in 1965) we quickly found that while some of the available textbooks were clearly the product of a basic philosophy of teaching akin to our own, they were yet in general inadequate for our purposes. We held that education for management is essentially creating the ability to analyse business problems to formulate viable solutions in a changing environment and to develop skills in bringing innovation about. It requires a comprehensive programme of studies depending on mathematical competence, financial knowledge, marketing expertise, a comprehension of the economic, technological and social forces working upon business, and a systematic study of individual and group behaviour. We found that the shortcoming of existing texts in following these aims was the lack of a sufficiently broad and rigorous, yet *consistent* coverage of materials. Some areas of teaching are, of course, better served than others, and this largely accounts for the sequence in which titles for the series are prepared— the most urgent needs, as we saw them, being met first. Thus we aimed on the one hand to produce texts suitable for adoption in the quantitative areas of a two-year Masters' programme and on the other, high-level and intellectually satisfying expository texts, again in the quantitative areas, suitable for adoption in post-experience programmes.

Partly also, our experience, and thus our perception of the functions textbooks should serve, has been formed by a critical but exceptionally constructive student audience. Many University teaching institutions are going through what is often felt to be a traumatic period of rising, and more articulate customers' demand for clear objectives in courses and relevant and economically presented material. As a new institution we have lived with the need continually to query and adjust objectives and methods right from the beginning. Students can no longer be called an 'audience', rather they are active participants in the teaching process. This is reflected in the textbooks in the series in that each results from practical classroom trials; participants on courses have contributed significantly to the final shape the texts take.

The teaching task in a graduate business school has, however, been shown to be inseparable from other functions an academic performs—notably consultancy and research. Good teaching depends on the stimuli of a constant challenge to make the material taught capable of application, and of contributing to the development of knowledge in the basic disciplines. An active professional life, in both these senses, complements the teaching work in such a school. Much of the writing arising out of such activities of course finds its way into the traditional avenues of publication. But some is more appropriately published in book form. The series thus will cover research monographs, collections of papers and proceedings of conferences of immediate interest to graduates in industry, and case materials.

M. E. BEESLEY

# Preface

Volume I is a detailed review of the available search procedures for solving optimization problems. It starts with a general definition of optimization, and then clarifies each of the mathematical structures appropriate to the various procedures. Each of the techniques is examined in detail and illustrated by numerical and graphical examples. The final chapter indicates the flexibility which is available for the mathematical definition of a problem. The first volume provides the tools which are needed for the applications discussed in volume II.

# 1 The study of optimization

## 1.1 Mathematics and practitioners

As optimization techniques have developed, a gap has arisen between the people devising the methods and the people who actually need to use them. Research into methods is necessarily long-term and located usually in academic establishments; whereas the application of an optimization technique, normally in an industrial environment, has to be justified financially in the short-term. The gap is probably inevitable; but there is no need for textbooks to reflect it. Teaching of optimization techniques separately from their connection with applications is pointless. How, otherwise, does a student know what to do with his skills? Volume I of this book gives a detailed exposition of the techniques and volume II investigates a wide range of applications to which the techniques apply.

The stress in many courses on optimization techniques is placed on the mathematics of the methods. Whole courses are devoted to individual techniques. Undoubtedly the mathematics is difficult and sufficiently taxing intellectually to be a discipline in its own right. But often there is a hint in the academic texts that the industrial practitioner was never meant to understand the ideas. A book on control theory which begins 'The process is considered to behave as a set of first order differential equations' will probably be meaningless to the chemical plant manager. Equally a monograph on time series analysis which immediately jumps into infinite vector spaces is not obviously meant to help the sales forecaster. It is highly desirable that the training in optimization should be appropriate for the potential practitioners as well as the theorists.

But there are difficulties. There are real grounds for accepting the notion of 'the mathematical blockage': that is the fence which divides those who can think mathematically from those who are unwilling or claim they cannot. It can be a real hurdle in a course on optimization in a Business School where the students come from varied academic backgrounds. However, not very much is demanded here. All that is required is an elementary knowledge of calculus, matrix algebra and probability distributions. It would also be helpful to know the form of simple functions for sketching graphs. These skills may be available at varying levels of confidence in different students. Were it not for our penchant for too early specialization in schools, they could well be routine to all. But there is usually a proportion of students for whom the statement 'let $x(t)$ be the number of cars stopped at the traffic lights at time $t$' remains a mystery. Each time $x(t)$ crops up they will have to grind laboriously through the process of visualizing cars lined up at a traffic intersection. The obstacles to quantitative fluency are often an illusion based on doubt and mixed with a trace of obstinacy. There is no intellectual breakthrough when a student begins to work with symbols like $x(t)$ isolated from their physical meaning. It is simply a matter of accepting a shorthand style.

The first two chapters of the book on basic concepts attempt to mitigate the language difficulty. The first one lists the basic mathematical concepts which it is assumed are known, and the subsequent chapter defines the general nature of an optimization problem. The notion of problem variables, objective functions and constraints which are introduced here will recur repeatedly throughout the book. A big effort has been made at keeping to the symbols presented in these chapters consistently throughout the text. It is worthwhile even for a mathematically trained student to read through these chapters to agree on the terms.

The next part of the book is devoted to an exposition of the full range of mathematical optimization techniques. It is hoped that the study of these techniques will prove really enjoyable. It is fascinating to see how the different ideas have been proposed for the various problem structures. The list of techniques forms a catalogue of schemes appropriate for the different forms of the variables and functions. Care has been taken to minimize the difficulties of the presentation. Graphical diagrams accompany the functional forms, and the common sense in the methods is clearly pointed out by geometrical arguments. For, however seldom it may be displayed in the textbooks, useful mathematical techniques often have a common sense and intuitive basis. The mathematical derivation of formulae and equations is clarified and supported by the physical interpretation being written down beside the expressions. Although this often implies a duplication, the option of a double reading in a tutorial text can only be an advantage. The policy throughout the text has been to maximize the facility for seeing what is 'going on'.

The more difficult sections and the more complex exercises have been marked with an asterisk, and the proofs of the theorems are held over for appendices.

All readers are strongly recommended to read through these sections. The interested practitioner will get just as much exhilaration from understanding a mathematical proof or complex derivation as will the student of the theory through realizing the successes of the methods in real applications. Worthwhile achievements do not usually arrive without some difficulties being encountered and overcome.

## 1.2 Applying optimization methods

Despite the tendency to teach the methods only, the organization of the practical problem and the selection of an optimization technique possess just as many intellectual difficulties and much more agony than the understanding of the methods. The successful application of an optimization technique requires, first, the recognition and formulation of an optimization problem, and secondly, the use of the answers provided by the solution.

The recognition of an optimization problem is not too difficult We need a set of choices, and the ability to control these choices, and a rule for deciding which of two alternative choices is preferrred. The optimization expert can visualize these components of an optimization problem occurring in numerous situations, but it is another thing to go out and find actual data. Optimization problems are not visibly lying around on the factory floor, although there may be plenty there. The factory planner, the manager or the process engineer seldom think of their objectives in terms of quantifiable formulae. Constraints and special restrictions on choices are often overlooked in the initial definition where they are established as a routine. Eventually an optimization study emerges by persuasion: it is a mixture of standing back from the situation and getting involved in the process where the optimization problem lies. Teaching and the reading of textbooks is much too straightforward and clean: they rarely communicate the confusion and messiness of actually doing something. Throughout the applications chapters of this text we will always start by describing the physical problem often using real data, and it is hoped that this practice and the breadth of the applications covered will help to give a grasp of the problem identification phase.

The mathematical definition follows the problem recognition, and volume II of the book considers the issues involved. Problem formulation is a subject much under-nurtured in education: it is always left to experience to produce it as a by-product. Yet is is probably the most valuable skill of all. Problem formulation in optimization is the specification of the mathematical form for the variables, the objective function, and constraints. It is customary to think of unique best answers to mathematical problems, but this is far from the case with problem formulation. A number of mathematical formulations of a problem may be equally valid depending on the point of view. The choice gives scope for imagination on the one hand and doubt on the other, both ingredients of realistic investigations. The detailed solutions to the applications exercises which are provided at the end of the book are only typical answers, and students providing alternative arguments which are nevertheless correct, should be glad to have seen a different line of attack.

In the end, optimization techniques have to be justified by some external criterion which is usually commercial. The question of whether the application of the optimization technique will increase profits or reduce costs has got to be asked and answered. The answer must often be an estimate. One can point to successful applications: blending grades in a refinery, scheduling an airline fleet, optimization in chemical process control, paper trim wastage minimization, transformer design, rubber compounding design, traffic flow control, and so on. These and many others are all published success stories and they are fully reviewed in the applications chapters. The emphasis has been placed on applications which have actually succeeded practically, not just educational case studies or hypothetical examples. But the successes are seldom quantified in terms of thousands of pounds saved per annum. This is not just because it is difficult to measure all the effects of the introduction of a new system. The measure of saving is often kept from competitors. But if commercial companies publish the fact that they are using optimization methods, it may be surmised that their published adventures were not flops.

Even when an optimization study is justified, and the problem is formulated and solved, the implementation of the answers may be a delicate operation. The first use of an optimization technique in a factory implies a new proposal and a new system and this will often be resisted. For example, unless one is careful, a design optimization study can be mistaken as the replacement of the designer rather than the provision of a calculating tool to offer a flexible design policy and the elimination of much tedious arithmetic. It is not possible to expand in this text on implementation policy. Successful implementation depends on taking care with the formulation, incorporating the right people into the optimization application at the right time, and having stamina.

Finally, one must have faith and common sense. A potential for improvement must exist in a system and one must intend to exploit it. Equally, the environment must be right and the supporting equipment available. It is no use hoping to apply an optimization method if the data cannot be obtained or is unreliable. Furthermore, the system must be reasonably stable: by the time the problem is solved and the technique provided, the original application must still exist. There is no point in proposing a circuit design optimization method if the technology it relates to is six months out of date. Probably it is necessary to have access to a computer and a programming system. These are all essential conditions for successful applications although we shall not study them further here. The following chapters aim to teach a student how to recognize, formulate and solve an optimization problem, and if this is achieved the book has been worth while. The opportunities for these skills are enormous. The field of industrial optimization is wide open. Most problems have still to be solved.

# 2 Mathematical background

The sections below list the basic mathematical concepts which are assumed to be known for the subsequent chapters. The statements are very brief and are not in any way intended as tuition in the concepts; they are rather to be used as a reminder and perhaps a clarification of the definitions.

## 2.1 Variables and functions

Variables are the basic elements of mathematical analysis. A variable is any quantifiable entity, for example the temperature of a furnace, or the number of vehicles owned by a distributor. Variables may vary continuously in a range or be restricted to a set of discrete values. The temperature in a furnace might vary continuously in the range 200°C to 700°C, but the number of vehicles owned by the distributor would be one of a discrete set of positive integers 0, 1, 2, .... A variable is typically denoted by the letter $x$ which identifies both the name of the variable and its value. If a problem is defined in terms of a number of similar variables, such as the ingredients in a chemical mix, the quantities may all be denoted by the letter $x$ and distinguished by subscripts: thus $N$ variables would be denoted by $x_1, x_2, ..., x_N$.

It is assumed that the following basic notation for the operations on variables and the relationships between them is well known. Exponentiation, $x^a$; the logarithm of $x$, $\log x$; modulus of $x$, $|x|$; the factorial of an integer $n$, $n!$; the sigma symbol, $\Sigma$; the inequality symbols $<, \leqq, >, \geqq$.

When the values of two variables are connected the relationship is expressed by a functional equation. Normally one variable called the dependent variable is calculated by performing a series of operations on the other variable called the independent variable. For example, the congestion level in a city may depend on the time of day. If $y$ is the dependent variable and $x$ is the independent variable, the general relationship is written as

$$y = f(x),$$

where $f$ is a symbol denoting the formula to be used to calculate $y$ given the value of $x$. For example the function $y = 4x^2 - 5x + 3$ enables $y$ to be calculated when $x$ is given. The dependent variable $y$ may depend on a number of variables, in which case the relationship would be written as

$$y = f(x_1, x_2, ..., x_N).$$

For example the yield from a chemical plant may depend on the rate of input of a number of raw materials.

The function relationship should not be confused with the solution of equations. We write the problem of solving an equation as

$$f(x) = 0.$$

This can be misleading as $f(x)$ does not always equal zero. It would be more accurate to state this problem as: find the values of $x$ for which $f(x)$ is zero. The values of $x$ for which the equation is satisfied are called the roots of the equation.

A typical inequality relationship is:

$$f(x) \geqq 0.$$

Again this may be ambiguous. The relation can be taken to mean that $f(x)$ is always positive. But like equation solving we may wish to find the values of $x$ for which $f(x) \geqq 0$. $f(x) = 1 + x^2$ is always positive, but $f(x) = 1 + x$ is only positive for $x > -1$. The inequality relationship must be interpreted in the context in which it is written.

## 2.2  Differentiation and integration of functions

The gradient of a function of a continuous variable at a point is evaluated by calculating the derivative, and the process of calculating it is called differentiation. The derivative of $f(x)$ is denoted by $\dfrac{df}{dx}$ and it is calculated by the limiting process

$$\frac{df}{dx} = \lim_{h \to 0} \frac{f(x+h) - f(x)}{h}.$$

When the derivative is to be evaluated at a particular point, say $x = a$, this is written as

$$\left[\frac{df}{dx}\right]_{x=a}.$$

The derivative of a function in $N$ variables $F(x_1, x_2, ..., x_N)$ in the direction of the variable $x_i$ is denoted by $\frac{\partial F}{\partial x_i}$ and this is calculated similarly as

$$\frac{\partial F}{\partial x_i} = \lim_{h \to 0} \frac{f(x_1, x_2, ..., x_i+h, ..., x_N) - f(x_1, x_2, ..., x_N)}{h}.$$

These quantities $\frac{df}{dx}$ and $\frac{\partial F}{\partial x_i}$ are called first-order derivatives. The derivatives of a higher order such as $\frac{d^2f}{dx^2}, \frac{\partial^2 F}{\partial x_i^2}, \frac{\partial^2 F}{\partial x_1 \partial x_2}$, etc., may be calculated by repeated application of the same procedure. Standard rules are used for the differentiation of products or quotients of functions, or functions of functions. If $f(x)$ and $g(x)$ are two functions of $x$, the derivative of their product is expressed as

$$\frac{d}{dx}(f(x)g(x)) = f(x)\frac{dg}{dx} + g(x)\frac{df}{dx}.$$

The derivative of the quotient is expressed as

$$\frac{d}{dx}(f(x)/g(x)) = \left(g(x)\frac{df}{dx} - f(x)\frac{dg}{dx}\right)/(g(x))^2.$$

If $y = g(x)$, the derivative of the function $f(y)$ with respect to $x$ is

$$\frac{dg}{dx}\left[\frac{df}{dy}\right]_{y=g(x)}.$$

A function in one variable is said to be monotonic increasing if its first derivative is always greater than zero, i.e. $\frac{df}{dx} \geq 0$, and to be monotonic decreasing if $\frac{df}{dx} \leq 0$.

The derivatives of a function may be used to obtain a useful relationship between two neighbouring points of a continuous function. If $a$ and $b$ are two points we may express $f(b)$ in terms of the value of $f(a)$ and the derivatives at $x = a$ as:

$$f(b) = f(a) + (b-a)\left[\frac{df}{dx}\right]_{x=a} + \frac{(b-a)^2}{2}\left[\frac{d^2f}{dx^2}\right]_{x=a}$$
$$+ ... + \frac{(b-a)^{n-1}}{(n-1)!}\left[\frac{d^{n-1}f}{dx^{n-1}}\right]_{x=a} + \frac{(b-a)^n}{n!}\left[\frac{d^nf}{dx^n}\right]_{x=a+c(b-c)}$$

where $0 \leq c \leq 1$. This is called the Taylor series expansion. Clearly if $b$ is close to $a$, the terms $\left(\dfrac{b-a}{n!}\right)^n$ become very small, and we may obtain a good approximation with the first three terms as

$$f(b) = f(a) + (b-a)\left[\frac{df}{dx}\right]_{x=a} + \frac{(b-a)^2}{2}\left[\frac{d^2f}{dx^2}\right]_{x=a}.$$

The Taylor series expansion may also be applied to a function in several variables to give an approximate relationship. If $B$ is the point $(b_1, b_2, ..., b_N)$, and $A$ is the point $(a_1, a_2, ..., a_N)$ and $X$ is the point $(x_1, x_2, ..., x_N)$, the first three terms in the Taylor series expansion about the point $A$ are

$$f(B) = f(A) + \sum_{i=1}^{N}(b_i - a_i)\left[\frac{\partial f}{\partial x_i}\right]_{X=A} + \tfrac{1}{2}\sum_{i=1}^{N}\sum_{j=1}^{N}(b_i - a_i)(b_j - a_j)\left[\frac{\partial^2 f}{\partial x_i \partial x_j}\right]_{X=A}.$$

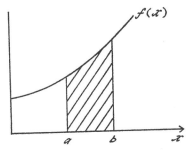

Fig. 2.1

The area enclosed by the function $f(x)$, the interval $(a \leq x \leq b)$ of the $x$-axis, and the ordinates at $x = a$ and $x = b$ is called the integral of $f(x)$ between $a$ and $b$ and the process of calculating it is called integration. The area is shown shaded in Fig. 2.1.

The integral symbol for the shaded area is denoted by $\displaystyle\int_a^b$, where $a$ and $b$ are called the limits of integration. The integral is calculated by the reverse procedure to differentiation. Denoting by $G(c)$ the area from $x = 0$ to $x = c$, the area is

$$G(b) - G(a) = \int_a^b f(x)dx$$

where the function $G(x)$ is determined by the relation

$$\frac{dG}{dx} = f(x).$$

## 2.3 Vectors

Vector notation provides a shorthand means of denoting a point in many dimensional space. The idea of representing a set of variables by the single letter

$$X = (x_1, x_2, ..., x_N)$$

has already been noted. The quantity $X$ is called a $N$-dimension vector and it refers to an ordered list of variables. For example if $N = 3$, the vector

$$X = (2, 4, 7)$$

implies $x_1 = 2$, $x_2 = 4$, $x_3 = 7$. A one-dimensional vector is called a scalar. For purposes of notation we will normally use capital letters for vectors and lower-case letters for scalars.

The length of the vector $X = (x_1, x_2, ..., x_N)$ is defined as

$$L(X) = \left\{ \sum_{i=1}^{N} x_i^2 \right\}^{\frac{1}{2}}.$$

This corresponds to the length of the line joining the origin to the point $(x_1, x_2. ..., x_N)$.

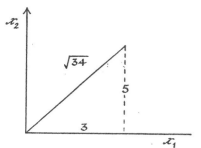

Fig. 2.2

In two dimensions Pythagoras' theorem states that the point (3, 5) is distant $\sqrt{(9+25)} = \sqrt{34}$ from the origin (Fig. 2.2).

The vectors have been written out above as rows of components. However, for the multiplication of vectors it is necessary to distinguish two ways of writing vectors and these will be called row and column vectors. A row vector has the components written out in a row and a column vector has the components written downwards in a column. If $X$ is an $N$ dimensional vector with components $x_1, x_2, ..., x_N$, it is written as a row vector as

$$(x_1, x_2, ..., x_N),$$

whereas, as a column vector, it is written as

$$X = \begin{pmatrix} x_1 \\ x_2 \\ \vdots \\ x_N \end{pmatrix}.$$

Wherever it is necessary, the arrangement of the vector—row or column— will always be defined.

The shorthand notation becomes useful for operations on vectors. The operations can only be performed if the vectors have the same dimension. If $X$ and $Y$ denote two vectors of $N$ dimensions with components $x_i$ and $y_i$, $i = 1$ to $N$, their addition is

$$X + Y = (x_1 + y_1, x_2 + y_2, ..., x_N + y_N).$$

Vectors of the same dimension can be multiplied into one another to provide a scalar quantity or number. The row and column arrangements of vectors must be considered here. If $X$ is a row vector and $Y$ is a column vector their scalar product is written as $X.Y$ and it is calculated by multiplying the corresponding components and adding up the sum as:

$$X.Y = x_1.y_1 + x_2.y_2 + ... + x_N.y_N.$$

The effect of multiplying a column vector into a row vector is considered later in matrix multiplication.

Vectors may be multiplied by scalars. If the scalar quantity $a$ is multiplied into $X$ it produces a vector with all components multiplied by $a$:

$$a.X = (ax_1, ax_2, ..., ax_N).$$

The idea of linear dependence is important in linear problems. It is defined in a rather complex way. The set of vectors $X_1, X_2, ..., X_r$ where

$$X_i = (x_{i1}, x_{i2}, ..., x_{iN})$$

are linearly dependent if there exist scalars $a_1, a_2, ..., a_r$ not all zero such that

$$a_1 X_1 + a_2 X_2 + ... + a_r X_r = 0.$$

The significance of linear dependence is that if the $N$ vectors of coefficients of $N$ equations in $N$ variables are linearly dependent, the equations cannot be solved uniquely.

## 2.4 Definition of regions

A region $R$ in $N$ dimensional space is often defined by the set of points which satisfy a given set of equations or inequalities. If there are $L$ equations and $M$ inequalities, they may be written as

$$H_k(x_1, x_2, ..., x_N) = 0 \text{ for } k = 1 \text{ to } L$$

$$G_k(x_1, x_2, ..., x_N) \leq 0 \text{ for } k = 1 \text{ to } M.$$

The region $R$ is defined by the set of points $(x_1, x_2, ..., x_N)$ which satisfy these conditions. The region does not exist if there are no points which satisfy the conditions. Alternatively the region may be defined just by equalities or just by inequalities. For example, the region defined by the three inequalities $x_1 + 2x_2 \leq 4$, $x_1 \geq 0$, $x_2 \geq 0$ can be drawn out on a two-dimensional diagram as the shaded region shown in Fig. 2.3.

A region $R$ is convex if for any two points

$X = (x_1, x_2, ..., x_N)$ and $Y = (y_1, y_2, ..., y_N)$

in the region the point $Z$ is also in the region $R$ where $Z = (z_1, z_2, ..., z_N)$ is any point defined as

$Z = aX + (1-a)Y$ for $0 \leq a \leq 1$.

Geometrically this means that the straight line connecting any two points of a convex region belongs to the region. In the two-dimensional plane the area enclosed by a triangle or a circle is a convex region. Convexity is an important concept in optimization theory.

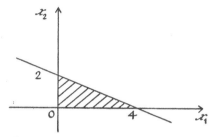

Fig. 2.3

## 2.5 Matrices

A matrix is a set of elements arranged in a rectangular array having a given number of rows and columns. The matrix $A$ of $M$ rows and $N$ columns with elements $a_{ij}$ is expressed as

$$A = \begin{bmatrix} a_{11} & a_{12} & ... & a_{1N} \\ a_{21} & a_{22} & ... & a_{2N} \\ . & . & . & . \\ a_{M1} & a_{M2} & ... & a_{MN} \end{bmatrix}.$$

The matrix $A$ is referred to as an $M$ 'times' $N$ matrix. A vector is a special case of a matrix, a column vector of $N$ components being an $N \times 1$ matrix and a row vector being a $1 \times N$ matrix.

Two matrices may be added together if they have exactly the same dimensions. If $A$ and $B$ are two matrices of dimensions $M \times N$, with elements $a_{ij}$ and $b_{ij}$, then the $(i, j)$th element of the matrix $(A+B)$ is $a_{ij} + b_{ij}$.

A matrix may be multiplied by a scalar. If the matrix $A$ is multiplied by the quantity $b$, the $(i, j)$th element of the resulting matrix has the value $b.a_{ij}$.

Two matrices may be multiplied together under special circumstances. The product of $A.B$ is achieved by multiplying the rows of $A$ into the columns of $B$ as in vector multiplication. The $(i, j)$th element—the element in the $i$th row and $j$th column—of the product is the multiplication of the $i$th row of $A$ into the $j$th column of $B$ giving

$$a_{i1}b_{1j} + a_{i2}b_{2j} + \ldots + a_{iN}b_{Nj}$$

and this quantity can be calculated only if $A$ has $N$ columns and $B$ has $N$ rows. Therefore if $A$ is an $M \times N$ matrix and $B$ is an $N \times P$ matrix, the product is an $M \times P$ matrix.

The transpose of a matrix is obtained by interchanging the rows and columns. The transpose of the matrix $A$ is written as $A'$. The $(i, j)$th element of the transpose $A'$ of the matrix $A$ is $a_{ji}$. In particular the transpose of a row vector is a column vector and vice versa.

The inverse of a matrix is defined only if the matrix is square, i.e. has an equal number of rows and columns. Furthermore it is only possible to find the inverse of a matrix $A$ if the vectors formed by the columns of $A$ form a linearly independent set. The inverse of $A$ is denoted by $A^{-1}$ where

$$A \cdot A^{-1} = I$$

the matrix $I$ being the identity matrix with unit values in the diagonal and zeros otherwise.

## 2.6 Permutations

A permutation of the $N$ elements $e_1$, $e_2$, ..., $e_N$ is an ordered list of these elements. If $p_j$ denotes the element in $j$th position of the list, the permutation may be denoted by

$$[P] = [p_1, p_2, \ldots, p_N].$$

A permutation should not be confused with a vector in $N$ dimensional space. The magnitude of $p_j$ has no significance; it is only the position in the sequence which is of importance. In a vector the components must be numbers, whereas the permutation may refer to an ordering of non-numerical elements.

A permutation may be altered by changing the positions of some of the elements in the permutation. A permutation exchange $E$ is a specified means of altering the positions of some of the elements in a permutation. For example, an exchange may define the interchange of the second and fifth elements in a permutation of 6 elements. If the permutation of the 6 elements was

$$[e_2, e_5, e_1, e_6, e_3, e_4]$$

the exchanged permutation would be

$$[e_2, e_3, e_1, e_6, e_5, e_4].$$

## 2.7 Set notation

A set is a symbol denoting a collection of elements. A set is normally denoted by a capital letter, the elements being denoted by small letters. If the set $S$

contains an element $e$, this is denoted by the symbolic relation

$e \in S$

which is read as '$e$ belongs to $S$' or '$e$ is contained in $S$'.

Any subset of the elements in $S$ is included in $S$. If $T$ is a subset belonging to $S$ we write

$T \subset S$

which is read as '$T$ is included in $S$'.

## 2.8 Definition of a network

A network is defined by a set $P$ of node points (or simply nodes) and a set $C$ of connections between pairs of node points. If there are $N$ nodes, they are

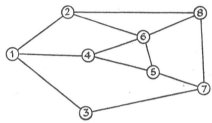

Fig. 2.4

usually identified by the integers 1 to $N$. The connections are denoted by a pair of numbers $(a, b)$ where $(a, b) \in C$. All this information may be contained in a network matrix $V$ where

$V(i, j) = 1$ if $(i, j) \in C$

$\qquad = 0$ otherwise.

The connections may have a measure of distance associated with them. For example, $d(i, j)$ may denote the distance between nodes $i$ and $j$.

A network is usually displayed graphically. Fig. 2.4 illustrates a network of 8 nodes with 12 connections.

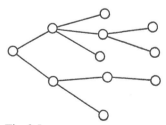

Fig. 2.5

A special form of network is a tree which consists of a root node from which all the other nodes can be reached by connections or branches from the root. Fig. 2.5 illustrates a tree.

## 2.9 Probability distributions

Many quantities are subject to some form of random variation. This random variation is described by a probability distribution. If the variable $x$ can assume the set of discrete values $x^{(1)}$, $x^{(2)}$, ..., $x^{(N)}$ we attach a probability measure $p(x^{(1)})$, $p(x^{(2)})$, ..., $p(x^{(N)})$, to each of these values, and state that Probability $(x = x^{(i)}) = p(x^{(i)})$.

The constraint $\sum_{i=1}^{N} p(x^{(i)}) = 1$ is imposed on the values $p(x^{(i)})$ as a scaling factor. If the variable may vary continuously over an interval $a \leq x \leq b$, we denote the probability of the variable falling in the small interval

$(x - \frac{1}{2}dx, x + \frac{1}{2}dx)$ by

$f(x)dx$.

This quantity should be thought of as the area under the curve $f(x)$ over the small interval of length $dx$. $f(x)$ is called the probability density function, and for it to be a valid density function we require a corresponding scaling factor

$$\int_a^b f(x)dx = 1.$$

Two important measures associated with random variables are the mean and variance. The mean, or average value, indicates a central point around which the distribution is scattered. The mean $\mu$ is calculated for the discrete case as

$$\mu = \sum_{i=1}^{N} x^{(i)}p(x^{(i)})$$

and for the continuous case as

$$\mu = \int_a^b xf(x)dx.$$

The variance gives an indication of the spread of the distribution. The variance $\sigma^2$ is calculated for the discrete case as

$$\sigma^2 = \sum_{i=1}^{N} (x^{(i)} - \mu)^2 p(x^{(i)})$$

and for the continuous case as

$$\sigma^2 = \int_a^b (x - \mu)^2 f(x)dx.$$

The quantity $\sigma$ called the standard deviation is often more useful as it is measured in the same units as the variable $x$.

In applications we often want to know what is the mean value of an expression $g(x)$ where the variable $x$ is subject to variation. The mean value or expected value of $g(x)$ is denoted by $E(g(x))$ and it is calculated in the discrete case as:

$$E(g(x)) = \sum_{i=1}^{N} g(x^{(i)}) \cdot p(x^{(i)})$$

and in the continuous case as

$$E(g(x)) = \int_{a}^{b} g(x) \cdot f(x)dx.$$

## Exercises on Chapter 2

**1** Sketch the functions $x^2, \dfrac{1}{x} + 2$ for $x \geq 0$.

**2** Differentiate the functions $f(x) = \left(\sqrt{x} - \dfrac{1}{\sqrt{x}}\right)^3$ and find the partial derivatives of $F(x_1, x_2) = \dfrac{x_1 x_2 e^{-x_1}}{1 + \log x_1 x_2}$.

**3** Express the Taylor series to two and three terms for the function $f(x) = x^3$ and obtain approximations for $f(x)$ at the values $x = 1 \cdot 1, 1 \cdot 25, 1 \cdot 5, 2 \cdot 0$, using the value of $f(x)$ and its derivatives at $x = 1$.

**4** Find the value of the integral

$$\int_{0}^{3} (x^2 + e^{-x})dx.$$

**5** Calculate the product of the vectors $X \cdot Y$ where
$X = (1, 3, 7, -1)$, $Y = (2, -2, 1, -3)$
and determine their lengths.

**6** Show that the vectors $X_1, X_2, X_3$ are linearly independent, where
$X_1 = (1, 2, 2)$
$X_2 = (3, -1, 0)$
$X_3 = (2, 0, 3)$.

**7** Draw the region in two dimensions defined by the inequalities
$-x_1^2 + x_2 \leq 0$
$x_1 + x_2 - 2 \leq 0$
$x_2 \geq 0$
$x_1 \geq 0$.

**8** By choosing two points in the region defined in Exercise 7, show that the region is not convex.

**9** Give where possible the values of $(A+B)$ and $A.B$ and their transposes for the matrices

(i)  $A = \begin{pmatrix} 1 & 2 & 3 \\ 1 & 3 & 6 \end{pmatrix}$   $B = \begin{pmatrix} 1 & 1 & 1 \\ 1 & 2 & 3 \end{pmatrix}$

(ii)  $A = \begin{bmatrix} 0 & 1 \\ 1 & 2 \\ 2 & 3 \\ 3 & 4 \end{bmatrix}$   $B = \begin{pmatrix} 1 & 1 & 1 & 1 \\ 1 & 3 & 5 & 7 \end{pmatrix}.$

**10** Find the inverse of the matrix $\begin{pmatrix} 3 & 5 \\ 8 & 10 \end{pmatrix}.$

**11** An adjacent exchange of a permutation is defined as the interchanging of the positions of two adjacent elements in the permutation. Write down all neighbouring exchanges of the permutation $[e_6, e_1, e_4, e_3, e_2, e_5]$.

**12** Write down the set of connections in the network specified by the following network connections matrix.

|   | 1 | 2 | 3 | 4 | 5 |
|---|---|---|---|---|---|
| 1 |   | 1 | 1 |   | 1 |
| 2 | 1 |   | 1 | 1 | 1 |
| 3 | 1 | 1 |   | 1 |   |
| 4 |   | 1 | 1 |   | 1 |
| 5 | 1 | 1 |   | 1 |   |

Draw out the network showing the nodes and connections.

**13** Find the mean and variance of the discrete and continuous distributions

(i)

| $x^{(i)}$ | 0 | 1 | 2 | 3 | 4 |
|---|---|---|---|---|---|
| $p(x^{(i)})$ | $\frac{3}{15}$ | $\frac{3}{15}$ | $\frac{2}{15}$ | $\frac{5}{15}$ | $\frac{2}{15}$ |

(ii) $f(x) = 6x(1-x)$ for $0 \leq x \leq 1$.

Also find the expected value of $e^{-x}$ if $x$ is distributed by the distribution $f(x) = 2x$ where $0 \leq x \leq 1$.

# 3 Optimization problems and methods

## 3.1 Basic components

All optimization problems consist of three basic components: the problem variables, the objective function and the constraints. The problem variables are the quantities which can be controlled, the objective function sets the aim for the system and evaluates the effectiveness of any control settings, and the constraints limit the ranges within which the controls can be varied. These three basic components will appear with monotonous repetition.

## 3.2 Problem variables

The first step in any optimization problem is the identification of the problem variables. These are the quantities whose values we wish to determine. For example in a chemical plant, the rate of flow of a feedstock input may be one of the variables which is to be controlled for an optimal reaction. Or in a machine allocation problem, the choice of machines for a factory job may be the control variable. The first of these variables is continuous, the second is discrete valued.

Typically, the problem variables will be denoted by the symbols

$x_1, x_2, ..., x_N,$

there being a total of $N$ variables. We will denote an assignment of values to the problem variables as

$X = (x_1, x_2, ..., x_N).$

This will also be referred to as the point $(x_1, x_2, ..., x_N)$.

### 3.3 Objective functions

The objective function is a formula for evaluating the cost, or profit, or return, etc., associated with an assignment of values to the problem variables. The objective function for the $N$ variables $x_1$, $x_2$, ..., $x_N$ will usually be denoted by

$F(x_1, x_2, ..., x_N)$.

We wish to determine the assignment of values which will minimize (or maximize) this function. The maxima and minima are sometimes called the extreme points of the function.

It makes no mathematical difference whether we are minimizing or maximizing $F$. If the problem requires the maximization of $F$, this can be converted

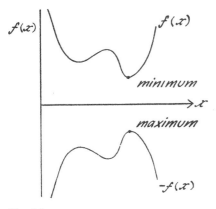

Fig. 3.1

into a minimization problem by considering $-F$. The writing of the minus sign creates the mirror image of the function and thus the maximum and minimum will be determined for the same problem variable values. Fig. 3.1 shows the effect of changing the sign for a function of one variable. In the development of the theory we will keep to the notion of minimizing a function except where it is more appropriate or conventional to deal with maximization problems.

*Example* 3.1
Given two problem variables $x_1$ and $x_2$, evaluate the objective function

$$F(x_1, x_2) = \tfrac{1}{2}x_1 + 2\frac{x_2}{x_1}$$

for the values $x_1 = 1, 2, 3, 4$ and $x_2 = -1, 0, 1, 2$.

It will be seen that the maximum value occurs at the point (1, 2) and the minimum at (1, −1). If the function was written as $-F(x_1, x_2)$ the maximum and minimum would be reversed.

| | | $x_2$ | | |
|---|---|---|---|---|
| | −1 | 0 | 1 | 2 |
| 1 | −1·5 | 0·5 | 2·5 | 4·5 |
| 2 | 0 | 1·0 | 2·0 | 3·0 |
| $x_1$  3 | 0·83 | 1·5 | 2·17 | 2·83 |
| 4 | 1·5 | 2·0 | 2·5 | 3·0 |

## 3.4  Constraints

In practical applications the problem variables are seldom allowed to range completely freely. For example the inlet flow rate in a pipe must lie between zero and the capacity of the pipe. The restrictions are expressed as constraints on the variables themselves or on functions of the variables, and these collectively define a region which will be called the feasible region. Any assignment $(x_1, x_2, ..., x_N)$ which satisfies the constraints is said to be a feasible solution.

The feasible region may take a variety of forms dependent on the type of constraints. It may be a set of discrete points or even a number of disjoint regions. The different kinds of constraint functions which will give rise to the various forms of region are considered in the following sections.

## 3.5  Equality constraints

Equality constraints have the mathematical form:

$$g(x_1, x_2, ..., x_N) = 0.$$

This is a very committing form of constraint. It expresses an exact relationship between the variables, and it can be used to eliminate one of the variables altogether.  For, if we know $x_1, x_2, ..., x_{N-1}$ the value of $x_N$ is automatically determined. To eliminate the $x_N$ variable the above equation can be expressed as:

$$x_N = h(x_1, x_2, ..., x_{N-1})$$

and this expression for $x_N$ can be substituted in the objective function, and any other constraints. However, obtaining the expression $h(x_1, x_2, .... x_{N-1})$ may be difficult or impossible, and, especially when some algebraic symmetry is preserved, it may be preferable to deal with the feasible region defined by the equality constraints. It may be noted here that the function $g$ expressing the constraint can include some partial derivatives.  This arises in process control applications where the rates of change over time of some of the variables are interrelated.

**B**

*Example* 3.2

Suppose the objective function

$$F(x_1, x_2) = 2x_1^2 + x_2^2$$

is to be minimized subject to the constraint

$$2x_1 + x_2 = 3.$$

The feasible region consists of all points on the line $2x_1 + x_2 = 3$, and the optimization problem is to find the minimum value of $2x_1^2 + x_2^2$ along this line.

Alternatively $x_2$ could be expressed in terms of $x_1$ as

$$x_2 = 3 - 2x_1.$$

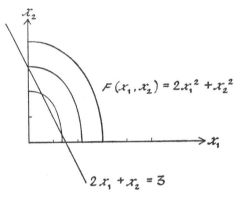

Fig. 3.2

When this is substituted in the objective function it gives a function in the single variable $x_1$:

$$F(x_1) = 2x_1^2 + (3 - 2x_1)^2$$

$$= 6x_1^2 - 12x_1 + 9.$$

We could now minimize $F(x_1)$ directly subject to no constraints.

## 3.6 Inequality constraints

Inequality constraints are expressed in the form

$$g(x_1, x_2, ..., x_N) \leq 0.$$

The equation $g(x_1, x_2, ..., x_N) = 0$ represents a surface in $N$ dimensions—a curve in a plane in 2 dimensions—and the constraint restricts the solution to lie on one side of the surface.

*Example* 3.3

A problem in two variables $x_1$ and $x_2$ has the following constraints:

$2x_1 - x_2 - 2 \geqq 0$

$x_1^2 - 4x_1 + x_2^2 \leqq 0$

$x_1 \geqq 0$

$x_2 \geqq 0.$

The feasible region can be drawn up as the shaded area shown in **Fig. 3.3**. It is determined by drawing up the curves and lines corresponding to the inequalities written as equations and investigating for a particular point which side of the curve or line the feasible region lies. The first constraint represents a line and the feasible region must lie to the right of the line as it is drawn in Fig. 3.3. The second constraint requires that the solution lies inside the circle.

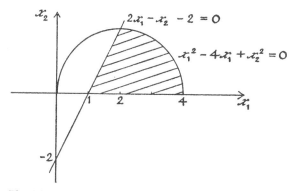

Fig. 3.3

The third and fourth constraints require the solution to lie in the first quadrant, i.e. for positive $x_1$ and $x_2$.

## 3.7 Discrete constraints

Discrete constraints require that individual variables take on one value of a set of possible values, for example the set of positive integers.

*Example* 3.4

A problem in two variables $x_1$, $x_2$ has three constraints

$x_1 \geqq 0$, $x_1$ integral

$5 \leqq x_2 \leqq 8$, $x_2$ integral

and $x_1 < x_2$.

The feasible region could then be drawn out as the set of possible points shown in Fig. 3.4.

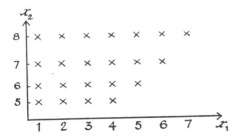

Fig. 3.4

## 3.8 Disjoint alternative regions

The feasible region may consist of a number of disjoint regions and it is required that the solution should be in one of the regions. Sometimes these constraints are referred to as 'either—or' constraints. If there are $m$ possible regions denoted by $R_1$, $R_2$, ..., $R_m$, we require that the solution should be in $R_1$ or $R_2$ or $R_3$ or ... or $R_m$. Each region $R_i$ will be defined by the constraint types already discussed.

*Example* 3.5

A problem in two variables $x_1$, $x_2$ has two alternative constraints. It is required that

either $x_1 \geqq x_2 + 2$

or    $x_2 \geqq x_1 + 5$.

The feasible region for this problem would be the areas shown as shaded in Fig. 3.5. At position $A$ the first constraint is being satisfied, although not the second. At position $C$ the second constraint is being satisfied but not the first.

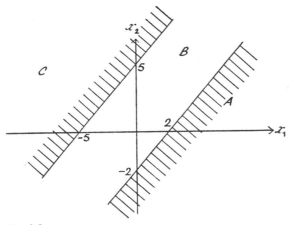

Fig. 3.5

At position $B$ neither constraint is being satisfied. Since only one or the other of the constraints must be satisfied, positions $A$ and $C$ are feasible points, but position $B$ is not feasible.

## 3.9 The general optimization problem

We have now discussed the meaning of the problem variables, the objective function and the various types of constraints which arise and we are in a position to propose the basic problem.

The general optimization problem is to determine the values of $N$ problem variables $x_1, x_2, ..., x_N$, so as to minimize an objective function $F(x_1, x_2, ..., x_N)$ subject to lying within a region $G$ which includes any constraints on the variables. The constraints define all feasible solutions, and we normally express the problem as the determination of values $(x_1, x_2, ..., x_N)$

to minimize $F(x_1, x_2, ..., x_N)$

subject to $(x_1, x_2, ..., x_N) \in G$.

An assignment of values which solves this problem is called an optimal solution. There are two kinds of optima which will now be distinguished.

## 3.10 Global optima

The global optimum is the proper solution to an optimization problem. It is the assignment of values which minimizes the function over the whole of the feasible region. The global minimum is therefore characterized by the following inequality. The solution $(x_1^*, x_2^*, ..., x_N^*)$ is the global minimum if

$$F(x_1^*, x_2^*, ..., x_N^*) \leqq F(x_1, x_2, ..., x_N)$$
for all $(x_1, x_2, ..., x_N) \in G$.

## 3.11 Local optima

In many problems it is not possible to find the global optimum as the time taken to conduct the search would be too long. We therefore define a local optimum as a solution which is optimum within a local region which is a subset of the feasible region. There may therefore be a number of local optima. The conditions for a solution being a local optimum are expressed similarly to the conditions for a global optimum. The solution $(\bar{x}_1, \bar{x}_2, ..., \bar{x}_N)$ is a local minimum if

$$F(\bar{x}_1, \bar{x}_2, ..., \bar{x}_N) \leqq F(x_1, x_2, ..., x_N)$$
for all $(x_1, x_2, ..., x_N) \in g$,

where $g \subset G$.

The local region may be determined by mathematical conditions or by an arbitrary rule. Continuous functions in two variables may have local 'bowls' in the surface which correspond to local optima. The tops of hills or mountains on a map are local maxima but only the highest peak on the whole map provides the global maximum. These local optima of continuous functions are defined mathematically. But in other situations, such as permutation problems, we will have to define the local optima more arbitrarily.

*Example 3.6*
An objective function in one variable has the form shown in Fig. 3.6, and the feasible region is $1 \leq x \leq 6$. The global minimum occurs at the point $p_4$ and

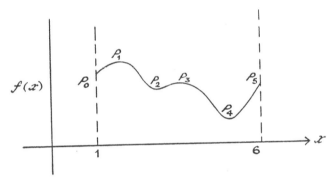

Fig. 3.6

the global maximum at $p_1$. There are local minima at $p_0$, $p_2$, $p_4$ and local maxima at $p_1$, $p_3$, $p_5$. These are optimal in the small region surrounding them.

## 3.12 The search for the optimum

The optimum solution is found by conducting a search over the feasible region. Usually this is organized in two stages. The first stage determines an initial feasible solution, and the second stage procedure improves the initial solution to obtain the optimum. Often an initial feasible solution is known from practical experience of the problem, and it is simply a question of employing the second stage. The second stage procedure is the main consideration of the optimization methods which will be presented.

The search procedure is usually a step by step calculation called an iterative method. Starting with the initial feasible solution we step from point to point, steadily improving the value of the objective function. At each step we decide the direction to move in and the distance to move before reassessing the situation. Suppose a problem has $N$ variables $x_1$, $x_2$, ..., $x_N$. Let the initial solution be denoted by

$$X^{(0)} = (x_1^{(0)}, x_2^{(0)}, ..., x_N^{(0)}).$$

The first step determines a new point $X^{(1)} = (x_1^{(1)}, x_2^{(1)}, ..., x_N^{(1)})$ for which

$$F(X^{(1)}) < F(X^{(0)})$$

(assuming the problem is a minimization). Then another step is taken to the point $X^{(2)}$ for which

$$F(X^{(2)}) < F(X^{(1)})$$

and so on. At each step the objective function is reduced. Fig. 3.7 illustrates how a series of steps is taken across the feasible region.

The success of the search procedure for finding the global optimum depends on the nature of the objective function and constraints. If the objective function is convex and the feasible region is also convex, we can start from any initial point, and move to the global optimum. But if the objective function

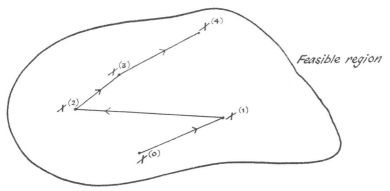

Fig. 3.7

is not convex or the feasible region is an awkward shape, we may end up in a local optimum depending on where the initial point lies and the stepping procedure we adopt. Not surprisingly, the ideal convex functions are referred to in optimization theory as 'well-behaved' functions.

## 3.13 Optimization methods

Optimization techniques propose the rules for moving from point to point in the track across the feasible region. They dictate how to choose a new direction and how far to move. The appropriate method depends on the form of the problem—the number of problem variables, whether they are continuous or discrete—and the form of the objective and constraint functions. To introduce the names of the optimization techniques and give an idea of the sorts of problems they will tackle, the titles of the methods are listed below together with a short statement about their capabilities.

*Calculus and Lagrange multipliers*
The calculus can be used on small problems of a few continuous variables with simple objective functions and no constraints. It can also be used when

equality constraints are present through the introduction of Lagrange multipliers. The calculus methods are elegant when they can be applied.

### Linear programming

Large problems in continuous variables can be solved by linear programming methods provided that the objective function is linear and the constraints are in the form of linear equalities and inequalities. The use of the method is pleasingly routine.

### Non-linear optimization

Non-linear optimization methods are concerned with handling the geometry of a non-linear objective function subject to non-linear inequality constraints which define the feasible region. The structure of these methods is more complex. Usually, it is only possible to determine a local optimum.

### Dynamic programming

Dynamic programming is a special search procedure for multi-stage decision processes. It solves a particular type of problem structure rather than a class of mathematical forms. Provided the right structure is present, constraints do not usually present many difficulties. The method skilfully reduces the amount of calculation which must be performed, given the right situations for its application. The success of the method depends on the correct formulation of the problem, and is often limited to comparatively small problems.

### Branch and bound methods

The branch and bound method (also called back-track programming) is a general approach to discrete optimization problems which aims to conduct an intelligent search through all possible solutions, cutting out large groups of possibilities early on in the search. It depends on being able to judge in advance which directions of search can be eliminated. Despite the generality of the idea, the method is limited to suitable small problems.

### Permutation procedures

This method deals with problems which are represented as permutations. It can find local optima of these problems, but the choice of how to define a local optimum must be decided for the particular application. This may be a useful freedom, but it also raises uncertainties about the quality of the answers being obtained.

### Heuristic techniques

Many industrial problems in which a fair amount is known about the structure of the problem, are solved by intuitive or heuristic approaches. Typically, a heuristic method will determine the value for each variable in turn, optimizing in a very short-term fashion. The computing time is quick but generally nothing is known about the quality of the answers.

For reference the characteristics of the problems which the various methods will solve are summarized in Table 3.1.

**Table 3.1**

| Method | Problem type | | | | |
|---|---|---|---|---|---|
| | Number of variables | Form of variables | Form of objective function | Form of constraints | Local or global optima |
| Calculus and Lagrange multipliers | few | continuous | differentiable | simple and differentiable equalities | local |
| Linear programming | many | continuous | linear | linear equalities and inequalities | global |
| Non-linear optimization | several | continuous | continuous and preferably convex | continuous functions and inequalities | local |
| Dynamic programming | several | preferably discrete | separable as multi-stage process | fairly general | global |
| Branch and bound methods | few | discrete | general | general | global |
| Permutation procedures | many | elements | general | general | specialized local |
| Heuristic techniques | many | discrete | well-structured | well-structured | unspeci-fied |

## Exercises on Chapter 3

**1** The objective function in two variables has the form:

$$F(x_1, x_2) = x_1^2 - 5x_1 + x_2^2 - 7x_2 + 10.$$

By tabulating the value of the objective function for integer values of $x_1$ and $x_2$ in the range

$$0 \leq x_1 \leq 4$$
$$0 \leq x_2 \leq 5$$

note where the minimum and maximum values lie.

**2** Sketch the feasible region for an optimization problem in two variables which has the following constraints:

$$(x_1 - 2)^2 \leqq x_2$$

$$x_1 - x_2 + 2 \geqq 0.$$

**3** Draw the feasible region in two variables for the constraints:

$$x_1 \geqq 0$$

$$x_2 \geqq 0$$

$$3x_1 + x_2 \leqq 9$$

$$x_1 + 2x_2 \leqq 8.$$

**4** Indicate on a diagram the feasible region for the two variable problem with constraints:

$$x_1^2 + x_2^2 = 16,$$

$\frac{1}{2}x_1$ integral.

**5** Sketch the feasible region for the alternative constraints in two variables,

either $(x_1 - 1)^2 + x_2^2 \leqq 1$

or $x_1 \geqq 3.$

**6** The objective function in one variable is expressed as:

$$F(x) = x^3 - 6x^2 + 9x + 4$$

and the variable $x$ is constrained to lie in the interval

$$-1 \leqq x \leqq 4.$$

Draw the curve and indicate all local minima and maxima.

*****7** For the feasible region of Exercise 3, if we add a further constraint

$$x_1 + x_2 \geqq a,$$

for what values of $a$:

(*i*) does the feasible region exist?

(*ii*) is the feasible region a single point?

(*iii*) is the feasible region non-existent?

# 4 Calculus and Lagrange multipliers

## 4.1 The achievement of the calculus

In the right circumstances the differential calculus can achieve a dramatic reduction in the search for the minimum or maximum of a function. The main conditions are that the function and its first derivative should be continuous, that the function should have a fairly simple form, and preferably be unconstrained. The property of continuous functions which the calculus uses is that the first derivative of the function must be zero at all the local maxima and minima. Geometrically this states that the gradient of the curve at the local maxima and minima is zero. This means that instead of posing the large problem of searching all possible values of $x$ to find the minimum value of $F(x)$, we can deal with the smaller problem of searching the comparatively few roots of

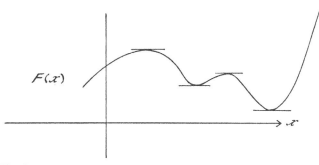

Fig. 4.1

the equation

$$\frac{dF}{dx} = 0$$

to obtain the minimum. For example the curve of Fig. 4.1 has four positions where $\frac{dF}{dx}$ is zero. We still have an equation-solving problem, but when this is simple the method of the calculus is an elegant optimization technique.

## 4.2  Restrictions on the use of the calculus

There are several important restrictions on the use of the calculus. First the derivative must be continuous. The calculus will not find the local maximum of the function shown in Fig. 4.2. At the maximum the value of the derivative

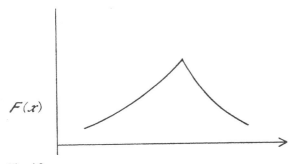

Fig. 4.2

'jumps' from being a positive value to a negative value and is therefore discontinuous.

Secondly, the calculus will not determine the local optima which occur at a constraint boundary. For example if the value of $x$ was constrained to lie in the interval $a \leq x \leq b$, the maximum and minimum values may be on the boundaries where the derivative is not zero. Fig. 4.3 illustrates a case. The

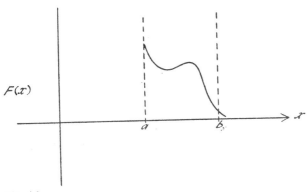

Fig. 4.3

calculus can therefore only be used on unconstrained functions with continuous first derivatives.

## 4.3 Problems in one variable

The procedure for problems in one variable is straightforward. We wish to find the minima of a function $F(x)$. We know for a continuous function that local minima occur where the gradient is zero. We therefore equate the derivative to zero by writing

$$\frac{dF}{dx} = 0$$

and solve the resulting equation.

Suppose $x = \bar{x}$ is a root of this equation. It may be a local maximum, minimum or point of inflexion of the function $F(x)$. A point of inflexion occurs when the function increases or decreases on both sides of the root $x = \bar{x}$. The root itself does not distinguish between these possibilities. A further investigation is needed. For $F(\bar{x})$ to be a local minimum, we require

$$F(x) \geq F(\bar{x}) \text{ for } \bar{x} - d \leq x \leq \bar{x} + d,$$

where $d$ is a small positive quantity. This states that $F(x)$ must be greater than $F(\bar{x})$ in the neighbourhood of the point $x = \bar{x}$. For a local maximum the same relation is written down with the first inequality reversed. $\bar{x}$ is a local maximum if

$$F(x) \leq F(\bar{x}) \text{ for } \bar{x} - d \leq x \leq \bar{x} + d.$$

Although the conditions are formally written down with the quantity $d$ present, it is not usually necessary to consider $d$ explicitly. An efficient means of deciding whether $x = \bar{x}$ is a local maximum or minimum is provided by the second derivative. The Taylor series expansion of the function $F(x)$ at a point $x$ close to $\bar{x}$ is

$$F(x) - F(\bar{x}) = (x - \bar{x})\left[\frac{dF}{dx}\right]_{x = \bar{x}} + \frac{(x - \bar{x})^2}{2}\left[\frac{d^2F}{dx^2}\right]_{x = \bar{x}} + \text{terms of order } (x - \bar{x})^3.$$

As $\left[\dfrac{dF}{dx}\right]_{x = \bar{x}} = 0$, since $\bar{x}$ is a root of the equation, this may be rewritten as

$$F(x) - F(\bar{x}) = \frac{(x - \bar{x})^2}{2}\left[\frac{d^2F}{dx^2}\right]_{x = \bar{x}} + \text{terms of order } (x - \bar{x})^3.$$

At points very close to $\bar{x}$ $\left(\text{provided }\left[\dfrac{d^2F}{dx^2}\right]_{x = \bar{x}} \text{ is not zero}\right)$, the terms of order $(x - \bar{x})^3$ can be ignored and the sign of $F(x) - F(\bar{x})$ is determined by the sign of $\left[\dfrac{d^2F}{dx^2}\right]_{x\,=\,\bar{x}}$ as $(x-\bar{x})^2$ will necessarily be positive. Thus

$F(x)-F(\bar{x})>0$ if $\left[\dfrac{d^2F}{dx^2}\right]_{x\,=\,\bar{x}} >0$, giving a local minimum

and

$F(x)-F(\bar{x})<0$ if $\left[\dfrac{d^2F}{dx^2}\right]_{x\,=\,\bar{x}} <0$, giving a local maximum.

Therefore the sign of the second derivative will usually determine the character of the function at $x = \bar{x}$.

If $\left[\dfrac{d^2F}{dx^2}\right]_{x\,=\,\bar{x}} = 0$, this rule is indeterminate. The point $\bar{x}$ may still be a point of inflexion or a minimum or a maximum and it is necessary to return to the conditions stated earlier to decide which it is. The point is a local minimum if

$$\frac{dF}{dx} \leq 0 \text{ for } \bar{x}-d \leq x \leq \bar{x}$$

and

$$\frac{dF}{dx} \geq 0 \text{ for } \bar{x} \leq x \leq \bar{x}+d$$

where $d$ is a suitably chosen small quantity. Similar conditions, with the inequality signs reversed, apply for a local maximum. For a point of inflexion we require that either $\dfrac{dF}{dx} \leq 0$ or $\dfrac{dF}{dx} \geq 0$ on both sides of $x = \bar{x}$.

*Example 4.1*

Find the local minima of the function

$2x^3 - 15x^2 + 24x + 12$ in the interval $(0, 7)$.

Differentiating and equating to zero:

$6x^2 - 30x + 24 = 0$, or

$6(x-1)(x-4) = 0$,

giving roots $x = 1$ and $4$.

Now $\dfrac{d^2F}{dx^2} = 12x - 30$ which is $<0$ for $x = 1$, and $>0$ for $x = 4$. Therefore $x = 4$ is a local minimum and $x = 1$ is a local maximum. If we consider points on the boundary as well, $x = 0$ is a local minimum and $x = 7$ is a local maximum, but the calculus does not provide these solutions.

## 4.4 Unconstrained problems in two or more variables

The differential calculus can equally be applied to unconstrained problems in several continuous variables although the considerations of what constitutes a local minimum or maximum becomes more complex. Intuitively, a local minimum or maximum of the function $F(x_1, x_2, ..., x_N)$ occurs where the tangent hyper-plane to the function $F$ does not increase in any direction. We therefore require that the partial derivatives in any direction $x_i$ should be zero, i.e.

$$\frac{\partial F}{\partial x_i} = 0 \text{ for } i = 1 \text{ to } N.$$

However, just as the first derivative of a function of one variable could not necessarily determine a local maximum or minimum, these conditions for many variables are not sufficient. A case is illustrated in the following example.

*Example 4.2*

Examine the nature of the function $F(x_1, x_2) = x_1 x_2$ at the origin. At the origin we have

$$\frac{\partial F}{\partial x_1} = x_2 = 0$$

$$\frac{\partial F}{\partial x_2} = x_1 = 0.$$

Therefore, both partial derivatives are zero. But consider a point very close to the origin with coordinates $(d_1, d_2)$. The magnitude of the function there will be $d_1 . d_2$. This will be positive if $d_1$ and $d_2$ are both positive or negative, but negative if one is positive and the other negative. Hence the function will increase in some directions from the origin but decrease in others as illustrated in Fig. 4.4. A point at which the partial derivatives are zero and at which a function behaves in this way increasing in some directions and decreasing in others is called a saddle point.

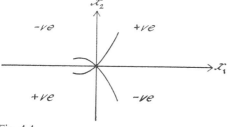

Fig. 4.4

We therefore need to find out if any point $(\bar{x}_1, \bar{x}_2, ..., \bar{x}_N)$ for which the partial derivatives are zero is a maximum, a minimum or a saddle point, and for this purpose it is necessary to know about the nature of the function in all directions from the point. Fortunately, we need not physically search all possible directions, as there are simple conditions which are usually adequate to determine the character of the point. The conditions are derived again by using Taylor's theorem. Because they are rather more complex, they will simply be stated here and proved in the Appendix (p. 37). For simplicity the conditions will be considered for two variables. The point $(x_1, x_2)$ for which

$$\frac{\partial F}{\partial x_1} = \frac{\partial F}{\partial x_2} = 0, \text{ is}$$

(*i*) a local minimum if

$$\frac{\partial^2 F}{\partial x_1^2}\frac{\partial^2 F}{\partial x_2^2} - \left(\frac{\partial^2 F}{\partial x_1 x_2}\right)^2 > 0$$

and

$$\frac{\partial^2 F}{\partial x_1^2} > 0 \text{ or } \frac{\partial^2 F}{\partial x_2^2} > 0;$$

(*ii*) a local maximum if

$$\frac{\partial^2 F}{\partial x_1^2}\frac{\partial^2 F}{\partial x_2^2} - \left(\frac{\partial^2 F}{\partial x_1 \partial x_2}\right)^2 > 0$$

and

$$\frac{\partial^2 F}{\partial x_1^2} < 0 \text{ or } \frac{\partial^2 F}{\partial x_2^2} < 0;$$

(*iii*) a saddle point if

$$\frac{\partial^2 F}{\partial x_1^2}\frac{\partial^2 F}{\partial x_2^2} - \left(\frac{\partial^2 F}{\partial x_1 \partial x_2}\right)^2 < 0;$$

(*iv*) undecided if

$$\frac{\partial^2 F}{\partial x_1^2}\frac{\partial^2 F}{\partial x_2^2} - \left(\frac{\partial^2 F}{\partial x_1 \partial x_2}\right)^2 = 0.$$

The last case is similar to the point of inflexion in a single variable problem.

*Example* 4.3
The function $F(x_1, x_2) = x_1^2 + x_2^2$ has a local minimum at the origin. For, at the origin, $\frac{\partial F}{\partial x_1} = 0, \frac{\partial F}{\partial x_2} = 0$ and $\frac{\partial^2 F}{\partial x_1^2} = 2, \frac{\partial^2 F}{\partial x_2^2} = 2, \frac{\partial^2 F}{\partial x_1 \partial x_2} = 0.$

Therefore $\frac{\partial^2 F}{\partial x_1^2}\frac{\partial^2 F}{\partial x_2^2} - \frac{\partial^2 F}{\partial x_1 \partial x_2} = 4 - 0 > 0$ and $\frac{\partial^2 F}{\partial x_1^2}, \frac{\partial^2 F}{\partial x_2^2} > 0.$

## 4.5 Equality constraints: Lagrange multipliers

The calculus procedures can also be used to solve problems in continuous variables when equality constraints are present by the introduction of additional variables called Lagrange multipliers. Suppose there are $M$ equality constraints which have the general form

$G_k(x_1, x_2, ..., x_N) = 0$, for $k = 1$ to $M(M \leqq N)$.

In principle these equations could be used to express $M$ of the variables in terms of the remaining $N - M$ variables, giving an unconstrained problem in $N - M$ variables. In practice it is often difficult or impossible to solve the equations. However, by means of the Lagrange multiplier method we can build the constraints directly into the objective function and avoid the initial equation solving problem. Furthermore, the method has the advantage of maintaining algebraic symmetry which can be helpful for solving the resulting derivative equations.

The Lagrange multipliers are auxiliary variables which are introduced into the problem. One variable or multiplier is required for each equation. Let us denote these variables by $y_1, y_2, ..., y_M$. Collectively they may be denoted by the vector $Y$. We now work with a new function $H(X, Y)$ consisting of the sum of the original objective function $F(X)$ and each constraint multiplied by its multiplier:

$$H(X, Y) = F(x_1, x_2, ..., x_N) + \sum_{k=1}^{M} y_k G_k(x_1, x_2, ..., x_N).$$

(It should be noted that the constraints $G_k(X)$ are written in the form $G_k(X) = 0$ before being added into the function $H(X, Y)$.) The function $H(X, Y)$ is now minimized by the previous calculus methods. Taking partial derivatives with respect to the $(N+M)$ variables

$x_1, x_2, ..., x_N$ and $y_1, y_2, ..., y_M$,

this provides $(N+M)$ equations for solution.

The Lagrange multiplier method is justified in the Appendix (p. 39) where it is shown that the unconstrained minimization of $H(X, Y)$ is equivalent to the minimization of $F(X)$ subject to $G_k(X) = 0$ for $k = 1$ to $M$.

*Example* 4.4
Find the maxima and minima of the function $x_1 x_2$ which lie on the circle

$x_1^2 + x_2^2 = 1$.

We construct the function

$H(x_1, x_2) = x_1 x_2 + y(x_1^2 + x_2^2 - 1)$

and differentiate this with respect to $x_1$, $x_2$ and $y$ and equate to zero providing three equations:

$$\frac{\partial H}{\partial x_1} = x_2 + 2yx_1 \text{ giving the equation } x_2 + 2yx_1 = 0,$$

$$\frac{\partial H}{\partial x_2} = x_1 + 2yx_2 \text{ giving the equation } x_1 + 2yx_2 = 0,$$

$$\frac{\partial H}{\partial y} = x_1^2 + x_2^2 - 1 \text{ giving the equation } x_1^2 + x_2^2 = 1.$$

To solve these equations a little algebraic manipulation is necessary. Multiplying the first equation by $x_2$ and the second equation by $x_1$ and adding gives the result

$$x_1^2 + x_2^2 + 4yx_1x_2 = 0.$$

Now using the third equation in this result gives

$$y = -\frac{1}{4x_1x_2}.$$

Substituting for $y$ in the first two equations provides the four solutions

$$x_1 = \pm \frac{1}{\sqrt{2}},$$

$$x_2 = \pm \frac{1}{\sqrt{2}}.$$

The nature of the function at the four points

$$\left(\frac{1}{\sqrt{2}}, \frac{1}{\sqrt{2}}\right), \left(\frac{-1}{\sqrt{2}}, \frac{-1}{\sqrt{2}}\right), \left(\frac{1}{\sqrt{2}}, \frac{-1}{\sqrt{2}}\right), \left(\frac{-1}{\sqrt{2}}, \frac{1}{\sqrt{2}}\right)$$

can be found by inspecting the function. At the first two points the function assumes a maximum value of $\frac{1}{2}$, and the other two points are minima where the function has the value of $-\frac{1}{2}$. These points are in fact global maxima and minima as there ar_ no boundary points.

It is worth while illustrating in this case the direct method in which one of the variables is eliminated. The equality constraint gives

$$x_2 = \pm (1 - x_1^2)^{\frac{1}{2}}.$$

Inserting this in the objective function (taking the positive sign),

$$F(x_1, x_2) = F(x_1, (1 - x_1^2)^{\frac{1}{2}}) = x_1(1 - x_1^2)^{\frac{1}{2}} = f(x_1)$$

a function of the single variable $x_1$. Differentiating

$$\frac{df}{dx_1} = (1 - x_1^2)^{-\frac{1}{2}}(1 - 2x_1^2)$$

which is zero for $x_1 = \pm \frac{1}{\sqrt{2}}.$

Substituting in the equation gives $x_2 = \pm \dfrac{1}{\sqrt{2}}$ and provides the same four points as before.

Generally the equation solving is not as simple as this and the Lagrange multiplier method must be used to reach a solution.

# Appendix

## *4.6 Sufficient conditions for local maxima and minima

We now prove the conditions stated in Section 4.4 for determining whether a point in a many variable function is a local maximum, minimum or a saddle point. The discussion is restricted to the case of two variables $x_1$ and $x_2$.

Let the conditions $\dfrac{\partial F}{\partial x_1} = \dfrac{\partial F}{\partial x_2} = 0$ hold at the point $(\bar{x}_1, \bar{x}_2)$. We wish to determine whether the function $F$ increases or decreases all round the point $(\bar{x}_1, \bar{x}_2)$ or whether it increases in some directions and decreases in others giving a saddle point.

Let $h$ and $k$ denote small displacements from the point $(\bar{x}_1, \bar{x}_2)$. The condition for a local maximum or minimum is that the quantity

$$Q(h, k) = F(\bar{x}_1 + h, \bar{x}_2 + k) - F(\bar{x}_1, \bar{x}_2)$$

should have the same sign for all sufficiently small values of $h$ and $k$. Taking the first three terms of the Taylor series for small $h$ and $k$,

$$F(\bar{x}_1 + h, \bar{x}_2 + k) = F(\bar{x}_1, \bar{x}_2) + h\left[\frac{\partial F}{\partial x_1}\right]_{\substack{x_1 = \bar{x}_1 \\ x_2 = \bar{x}_2}} + k\left[\frac{\partial F}{\partial x_2}\right]_{\substack{x_1 = \bar{x}_1 \\ x_2 = \bar{x}_2}}$$

$$+ \tfrac{1}{2}\left(h^2\left[\frac{\partial F}{\partial x_1^2}\right]_{\substack{x_1 = \bar{x}_1 \\ x_2 = \bar{x}_2}} + 2hk\left[\frac{\partial^2 F}{\partial x_1 \partial x_2}\right]_{\substack{x_1 = \bar{x}_1 \\ x_2 = \bar{x}_2}} + k^2\left[\frac{\partial^2 F}{\partial x_2^2}\right]_{\substack{x_1 = \bar{x}_1 \\ x_2 = \bar{x}_2}}\right) + E$$

where $E$ is an error term which may be ignored for sufficiently small $h$ and $k$.

Writing $a = \left[\dfrac{\partial^2 F}{\partial x^2}\right]_{\substack{x_1 = \bar{x}_1 \\ x_2 = \bar{x}_2}}$, $b = \left[\dfrac{\partial^2 F}{\partial x_1 \partial x_2}\right]_{\substack{x_1 = \bar{x}_1 \\ x_2 = \bar{x}_2}}$, $c = \left[\dfrac{\partial^2 F}{\partial x_2^2}\right]_{\substack{x_1 = \bar{x}_1 \\ x_2 = \bar{x}_2}}$,

the expression $Q(h, k)$ may be written as

$$Q(h, k) = \tfrac{1}{2}(ah^2 + 2bhk + ck^2)$$

since the first derivatives are zero. (In the exceptional case of $a$, $b$, $c$ all being zero the Taylor series must be examined to further terms.)

$Q(h, k)$ is called a quadratic form and it is said to be definite if it always has the same sign (positive or negative) regardless of the values of $h$ and $k$. It is said to be indefinite if it can assume either sign. The nature of $Q(h, k)$ is determined by establishing certain inequalities on the quantities $a$, $b$ and $c$.

$Q(h, k)$ may be rewritten as

$$Q(h, k) = \tfrac{1}{2}(ah^2 + 2bhk + ck^2)$$

$$= \frac{a}{2}\left[\left(h + \frac{b}{a}k\right)^2 + \left(\frac{ca - b^2}{a^2}\right)k^2\right]$$

and this expression is definite if $(ca - b^2) > 0$. It then has the same sign as $a$. It is indefinite if $(ca - b^2) < 0$ for it then may assume values of different signs for instance when $k = 0$ and $k = -\dfrac{a}{b}h$.

If $(ca - b^2) = 0$, there is a direction along which $Q(h, k)$ remains zero. This is the direction for which $\dfrac{h}{k} = -\dfrac{b}{a}$, i.e. the direction which has a gradient $-\dfrac{b}{a}$ with respect to the $x_1$-axis.

These conditions can now be rewritten in the original derivative forms. For a local minimum we require that $Q(h, k) > 0$. This means that $(ca - b^2) > 0$ and $a > 0$. (Note that $(ca - b^2) > 0$ and $a > 0$, implies $c > 0$.) The requirements for a minimum are

$$\frac{\partial^2 F}{\partial x_1^2}\frac{\partial^2 F}{\partial x_2^2} - \left(\frac{\partial^2 F}{\partial x_1 \partial x_2}\right)^2 > 0$$

and

$$\frac{\partial^2 F}{\partial x_1^2} > 0 \quad \text{or} \quad \frac{\partial^2 F}{\partial x_2^2} > 0.$$

For a local maximum the requirements are

$$\frac{\partial^2 F}{\partial x_1^2}\frac{\partial^2 F}{\partial x_2^2} - \left(\frac{\partial^2 F}{\partial x_1 \partial x_2}\right)^2 > 0$$

and

$$\frac{\partial^2 F}{\partial x_1^2} < 0 \quad \text{or} \quad \frac{\partial^2 F}{\partial x_2^2} < 0$$

For a saddle point

$$\frac{\partial^2 F}{\partial x_1^2}\frac{\partial^2 F}{\partial x_2^2} - \left(\frac{\partial^2 F}{\partial x_1 \partial x_2}\right)^2 < 0.$$

The case where $\dfrac{\partial^2 F}{\partial x_1^2}\dfrac{\partial^2 F}{\partial x_2^2} = \left(\dfrac{\partial^2 F}{\partial x_1 \partial x_2}\right)^2$ is undecided as we cannot form a definite opinion about whether $Q(h, k)$ does or does not change sign. It is necessary to investigate further terms of the Taylor series to establish the nature of the function at the point $(\bar{x}_1, \bar{x}_2)$.

Similar conditions for local maxima or minima can be established for problems in three or more variables. But it is then necessary to consider quadratic forms in more than two quantities. As this is quite straightforward requiring the use of determinants which can be found in any basic algebra text it will not be discussed further here.

## *4.7 Proof of the Lagrange multiplier method

The method of Lagrange multipliers for handling equality constraints is now proved for the general case. We wish to minimize (or maximize) the function

$$F(x_1, x_2, ..., x_N)$$

subject to the $M$ constraints

$$G_k(x_1, x_2, ..., x_N) = 0 \text{ for } k = 1, ..., M.$$

The equality constraints enable us to write $M$ of the variables, say

$$x_1, x_2, ..., x_M$$

as functions of the other variables as

$$x_j = h_j(x_{M+1}, x_{M+2}, ..., x_N).$$

For any point on the constraint the derivatives of the constraint function with respect to $x_r (r = M+1, M+2, ..., N)$ will be zero:

$$\frac{dG_k}{dx_r} = \sum_{j=1}^{M} \frac{\partial G_k}{\partial x_j} \cdot \frac{\partial h_j}{\partial x_r} + \frac{\partial G_k}{\partial x_r} = 0 \text{ for } k = 1, 2, ..., M.$$

Multiplying the $k$th constraint by any quantity $y_k$ and adding up for

$$k = 1, ..., M$$

gives the equation

$$\sum_{j=1}^{M} \frac{\partial h_j}{\partial x_r} \left[ \sum_{k=1}^{M} y_k \frac{\partial G_k}{\partial x_j} \right] + \sum_{k=1}^{M} y_k \frac{\partial G_k}{\partial x_r} = 0 \text{ for } r = M+1, M+2, ..., N.$$

Also a necessary condition for a maximum or minimum of $F$ is

$$\frac{\partial F}{\partial x_r} + \sum_{j=1}^{M} \frac{\partial F}{\partial x_j} \cdot \frac{\partial h_j}{\partial x_r} = 0 \text{ for } r = M+1, ..., N.$$

Adding up the last two sets of equations for any value of $r$, $M+1 \leqq r \leqq N$ gives

$$\frac{\partial F}{\partial x_r} + \sum_{k=1}^{M} y_k \frac{\partial G_k}{\partial x_r} + \sum_{j=1}^{M} \frac{\partial h_j}{\partial x_r} \left[ \frac{\partial F}{\partial x_j} + \sum_{k=1}^{M} y_k \frac{\partial G_k}{\partial x_j} \right] = 0.$$

Assuming $\dfrac{\partial G_k}{\partial x_j}$ is non-zero we may define $y_k$ for $k = 1, \ldots, M$ by the equations

$$\frac{\partial F}{\partial x_j} + \sum_{k=1}^{M} y_k \frac{\partial G_k}{\partial x_j} = 0, j = 1, \ldots, M$$

and the previous equation in $x_r$ implies

$$\frac{\partial F}{\partial x_j} + \sum_{k=1}^{M} y_k \frac{\partial G_k}{\partial x_j} = 0, j = M+1, M+2, \ldots, N.$$

These are the equations provided by the Lagrange multiplier method.

REFERENCES

Courant, R. 1937. *Differential and Integral Calculus*, Vols 1 and 2, sections on Maxima and Minima. Blackie, Glasgow. (Reprinted 1966.)

## Exercises on Chapter 4

**1** Given the function $f(x) = -x$ for $x \leq 0$ and $f(x) = x$ for $x > 0$, examine why the calculus fails to find the minimum of this function.

**2** Show that the calculus will not determine the minimum of the function

$f(x) = x^2 - 2x + 1$

subject to the constraint $x \geq 2$.

**3** Find the local maxima and minima of the function $x^3 - 18x^2 + 96x$ in the interval $(0, 9)$.

**4** Show that the function $x^3$ has no local maxima or minima.

**5** Show that the function $\dfrac{ax+b}{cx+d}$ has no local maxima or minima for any values of $a$, $b$, $c$, or $d$.

**6** Investigate the nature of the following functions at the origin:

(*i*) $x_1^2 + x_1 x_2 + x_2^2 + x_1^3 + x_1^2 x_2 + x_2^3$.

(*ii*) $x_1^2 + 3x_1 x_2 + x_2^2 + x_1^3 + x_1^2 x_2 + x_2^3$.

(*iii*) $x_1^2 + 2x_1 x_2 + x_2^2 + x_1^4 + x_1^2 x_2^2 + x_2^4$.

**7** Which point on the sphere $x_1^2 + x_2^2 + x_3^2 = 1$ is at the greatest distance from the point $(1, 2, 3)$? The distance of the point $(x_1, x_2, x_3)$ from $(1, 2, 3)$ is given by the formula

$\{(x_1 - 1)^2 + (x_2 - 2)^2 + (x_3 - 3)^2\}^{\frac{1}{2}}$.

**\*8** Find the rectangle whose sides are parallel to the axes with the greatest perimeter inscribed in the ellipse

$$\frac{x_1^2}{a^2} + \frac{x_2^2}{b^2} = 1.$$

**\*9** Determine the nature of the optimum of the function

$$x_1 x_2 + \frac{1}{x_1} + \frac{2}{x_2}.$$

**\*10** Determine the maxima and minima of the function

$$(ax_1^2 + bx_2^2)e^{-(x_1^2 + x_2^2)} \qquad 0 < a < b.$$

**\*11** Show how the maximum of the function

$$a_1 x_1^2 + a_2 x_2^2 + a_3 x_3^2$$

subject to

$$x_1^2 + x_2^2 + x_3^2 = 1$$
$$b_1 x_1 + b_2 x_2 + b_3 x_3 = 0$$

can be determined.

**\*12** The sum of the length of twelve edges of a rectangular block is $a$. The sum of the area of the six faces is $a^2/25$. Calculate the length of the edges when the excess of the volume of the block over that of a cube whose edge equals the least edge of the block is greatest.

# 5 Linear programming

## 5.1 Linear functions

The method of linear programming has been the most potent force towards the practical use of optimization techniques. The method was the first technique which would handle optimization problems in a large number of variables, and its ready implementation in a computer program made its use a routine procedure. Its general efficiency has led to various proposals for converting non-linear problems into a linear framework; some of these will be examined in the next chapter.

Linear functions are the simplest way of combining variables, consisting simply of a sum of the variables multiplied by constant coefficients. Linear programming deals with linear objective functions and linear equality or inequality constraints. If there are $N$ continuous variables $x_1, x_2, ..., x_N$, the objective function to be maximized or minimized has the form

$$F(x_1, x_2, ..., x_N) = c_0 + \sum_{i=1}^{N} c_i x_i$$

where the $x_i$ are continuous variables, and the $c_i$ are constants. The constraints must be linear equalities or inequalities of the form

$$\sum_{j=1}^{N} a_{ij} x_j = b_i$$

$$\text{or} \sum_{j=1}^{N} a_{ij} x_j \leq b_i$$

$$\text{or} \sum_{j=1}^{N} a_{ij} x_j \geq b_i$$

where the $a_{ij}$ and $b_i$ are constant coefficients. We will not consider here the case where the $x_j$ variables must be discrete. Where this occurs the problem is called an integer linear programming problem. In a later chapter on branch and bound methods we will examine a method for solving the integer problem.

The computational procedure of linear programming known as the simplex method does not deal with the full range of constraints which have been expressed above. Instead, the method tackles a standard linear programming problem in which the objective function is to be maximized, the constraints are all equalities, and the variables must be non-negative. However, the problems of minimizing a linear function or handling negative valued variables or inequality constraints can readily be converted into the standard form. We will first present the standard form, and then describe how the conversion can be arranged.

## 5.2 The standard form

The standard linear programming problem is to determine the values of the variables $x_1$, $x_2$, ..., $x_N$ so as to maximize the function

$$c_0 + \sum_{j=i}^{N} c_j x_j$$

subject to the linear equalities

$$\sum_{j=1}^{N} a_{ij} x_j = b_i \text{ for } i = 1, ..., M$$

and the non-negativity constraints

$x_j \geqq 0$ for $j = 1, ..., N$.

It is assumed that the $b_i$ values are positive or zero.
(It is also assumed that $N \geqq M$. If $N = M$ the values of $x_i$ are determined uniquely by the equations. If $N < M$, the situation is 'over-determined' and it is no longer an optimization problem.)

## 5.3 Converting to the standard form

First, if the objective function is to be minimized, it can be converted into a maximization problem by multiplying the objective function by $-1$. Instead of minimizing

$$c_0 + \sum_{i=1}^{N} c_i x_i$$

we maximize $-c_0 - \sum_{i=1}^{N} c_i x_i$.

When the optimum variable values have been determined, the actual value of the objective function must again be multiplied by $-1$ to give it its correct meaning.

Secondly, we consider the conversion of the constraints. If a constraint is expressed such that $b_i < 0$ this can be redefined so that $b_i > 0$ by multiplying the whole equation by $-1$. This will not affect the system of equations.

If a constraint is a 'less than' condition, as

$$\sum_{j=1}^{N} a_{ij}x_j \leqq b_i$$

this can be converted into an equality condition by introducing a non-negative variable $s_i$ into the left-hand side of the inequality to give the equation

$$\sum_{j=1}^{N} a_{ij}x_j + s_i = b_i.$$

The variable $s_i$ is called a 'slack' variable as it takes up the slack in the inequality. $s_i$ is now a variable just like the $x_j$. Similarly if a constraint is a 'greater than' condition as

$$\sum_{j=1}^{N} a_{ij}x_j \geqq b_i$$

this can be converted into an equality condition by introducing a non-negative slack variable with a minus sign as:

$$\sum_{j=1}^{N} a_{ij}x_j - s_i = b_i.$$

Finally, if some variables are not constrained to be non-negative, then we can express these variables as a difference between two non-negative variables as

$$x_j = x_j^+ - x_j^-$$

where

$$x_j^+ \geqq 0,$$

$$x_j^- \geqq 0.$$

This is possible because any number can always be expressed as the difference between two non-negative numbers. For instance $3 = (+3)-(+0)$ or $-7 = (+0)-(+7)$. Substituting this expression into the objective function and constraints for those variables which are not restricted in sign, we again return to the standard form where all variables must be non-negative.

*Example 5.1*

Convert to the standard form:

Minimize $-x_1 + x_2 - x_3$

subject to   $x_1 - 3x_2 + 4x_3 = 5$

$\qquad\qquad x_1 - 2x_2 \qquad\ \leqq 3$

$\qquad\qquad 2x_2 - x_3 \qquad\ \geqq 4$

and   $\qquad x_1 \geqq 0,\ x_2 \quad \geqq 0.$

Multiply the objective function by $-1$ to make it a maximization problem. Add a slack variable $x_4$ to the second constraint, subtract a slack variable $x_5$ from the third constraint and write $x_3$ as $x_3^+ - x_3^-$ since $x_3$ is unconstrained. This gives the revised problem:

Maximize $x_1 - x_2 + x_3^+ - x_3^-$

subject to $x_1 - 3x_2 + 4x_3^+ - 4x_3^- \qquad\qquad = 5$

$\qquad\quad x_1 - 2x_2 \qquad\qquad\ + x_4 \quad = 3$

$\qquad\qquad\ 2x_2 - x_3^+ + x_3^- \qquad - x_5 = 4$

$\qquad\quad x_1, x_2, x_3^+, x_3^-, x_4, x_5 \qquad = 0.$

## 5.4 The ideas of the simplex method

The simplex method is based on some important ideas relating to linear functions. The description of the method involves these concepts and we will therefore introduce them first.

In the standard form the problem is to determine $x_1, x_2, \ldots, x_N$ to maximize a linear function subject to $M$ equality constraints, where $M \leq N$. Also the variable values must be non-negative. Now there is no reason, in advance, to suppose that there will be any special conditions imposed on the $x_i$ values at the optimum. However, the Fundamental Theorem (proved in the Appendix, p. 61) of linear programming shows that the optimum solution will occur at a point for which at most $M$ variables have positive values. It will usually be exactly $M$ variables. The remaining $(N - M)$ variables will be zero. We will call a set of $M$ variables a basis. This result offers great possibilities for constructing a computational method. It means that we could take various bases of $M$ variables, set the values of the remaining variables to zero, and solve the resulting $M$ equations in $M$ variables. These are called basic solutions. However, there are $N!/\{(N-M)!M!\}$ possible ways of selecting the set of variables and this can be a large number. There is a further difficulty. Not all basic solutions are feasible. The non-negative constraints on the $x_i$ variables may be violated. We therefore need to concentrate on basic feasible solutions.

The next feature to note is that the collection of feasible solutions constitutes a convex set. (This result is also proved in the Appendix, p. 63.) The consequence of convexity is that starting from any initial feasible point we can move from point to point steadily improving the objective function value and thus reach the global optimum. This opens up the possibility of moving from basis to basis steadily improving the objective function value until we reach the optimal basis. The means of doing this algebraically is to start with one basis, say the first $M$ variables $x_1, x_2, ..., x_r, ..., x_M$ and at the next stage to remove one variable, say $x_r$ and bring in $x_s$ to provide a new basis. This is the procedure of the simplex method. These ideas can be pictured geometrically as shown in the following example.

*Example 5.2*

Suppose we have a simple two variable problem:

Maximize $x_1 + x_2$

subject to $2x_1 + 3x_2 \leqq 6$

$$2x_1 + x_2 \leqq 4$$

$$x_1, x_2 \geqq 0.$$

The inequality constraints are converted into equalities through slack variables $x_3$ and $x_4$ defining the pair of equations

$$2x_1 + 3x_2 + x_3 = 6$$

$$2x_1 + x_2 + x_4 = 4.$$

As there are two equations, there are two variables in a basic solution. A total of four variables means that there are $\binom{4}{2} = 6$ ways of choosing the two variables and correspondingly 6 basic solutions. These basic solutions are marked on the diagram by labelling the points with the pair of variables in the basic solution. For instance if $x_2$ and $x_3$ are in the basic solution, we set $x_1$ and $x_4$ to zero and solve the above equations giving the point $F$ in the $x_1, x_2$ plane.

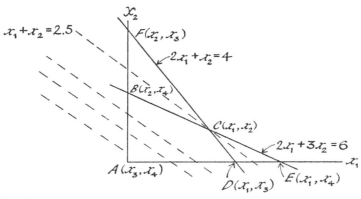

Fig. 5.1

It will be noticed that the set of feasible basic solutions corresponds to the points $A$, $B$, $C$ and $D$. For example the point $E$ with basic variables $x_1$ and

$x_4$ gives the solution $x_1 = 3$, $x_4 = -2$ which is not feasible, as $x_4$ must be positive. The feasible region $A$, $B$, $C$, $D$ is convex.

The value which the objective function attains in the feasible region is shown as a series of parallel lines across the diagram. Each line is determined for a particular value of $x_1 + x_2$. The objective function value steadily increases the further we move from the origin. By inspection its maximum value will occur at $C$, and this illustrates how the optimum will occur at a basic feasible solution. Furthermore, if we started from any of the basic solutions $A$, $B$ or $D$ we could move round the feasible region from basis to basis steadily increasing the value of the objective function. This is the scheme of the simplex method, although it is not necessary to adopt this approach in the simplest 2-variable linear programming problems as they can be solved by the kind of graphical arguments which have just been illustrated.

We will describe the structure of the simplex method in an unusual order. First we will show how to reduce the system of equations to the so-called canonical form. Later it will be seen what a helpful piece of algebra this is. Next we will assume that we have a basic feasible solution and examine whether it can or cannot be improved using the canonical form. Thirdly we will show how to obtain the next canonical form from the previous one. This is called a change of basis. Finally it will be shown how an initial basic feasible solution can be found presuming one is not available in advance.

It should be stressed that the canonical form presentation of the simplex procedure is not the best approach for computational purposes. But it is the presentation which enables the basis to basis movement to be grasped and followed through most easily. Some consideration will be given in Section 5.11 to the revised simplex procedure which is used in computer programs.

## 5.5 The canonical form

As we propose to work from basis to basis the first thing to know is how to solve the system of equations for any given basis. Let us recall the general form of the standard linear programming problem stated at the beginning of Section 5.2. If we write the objective function as an equation in which the value of the objective function is denoted by the variable $z$, the full system of equations including the objective function is arranged in the form:

$$a_{11}x_1 + a_{12}x_2 + \ldots + a_{1N}x_N = b_1$$
$$a_{21}x_1 + a_{22}x_2 + \ldots + a_{2N}x_N = b_2$$
$$\vdots$$
$$a_{MN}x_1 + a_{M1}x_2 + \ldots + a_{MN}x_N = b_M$$
$$c_1x_1 + c_2x_2 + \ldots + c_Nx_N = z - c_0.$$

We will take the basis variables as $x_1$, $x_2$, ..., $x_M$. The values of these variables are determined by converting the equations into their canonical form.

This is a manipulative procedure for isolating the basis variables. The process is equivalently achieved by inverting the matrix of coefficients of the basis variables; but it is easier to see what is going on by operating on the equations directly rather than using matrix notation.

The conversion process is as follows. The first equation is divided by $a_{11}$ (assumed $>0$) and it can then be used to express $x_1$ in terms of the other variables and substituted in the remaining equations. This provides the new system of equations:

$$x_1 + \bar{a}_{12}x_2 + \ldots + \bar{a}_{1N}x_N = \bar{b}_1$$
$$\bar{a}_{22}x_2 + \ldots + \bar{a}_{2N}x_N = \bar{b}_2$$
$$\vdots$$
$$\bar{a}_{M2}x_2 + \ldots + \bar{a}_{MN}x_N = \bar{b}_N$$
$$\bar{c}_2 x_2 + \ldots + \bar{c}_N x_N = z - \bar{c}_0$$

where $\bar{a}_{ij}$, $\bar{b}_i$, $\bar{c}_j$ are the revised constants. If we repeat this procedure to isolate $x_2, x_3, \ldots, x_M$ in turn and eliminate them from the remaining equations we obtain the final system of equations as:

$$x_1 \qquad + a'_{1, M+1}x_{M+1} + \ldots + a'_{1N}x_N = b'_1$$
$$x_2 \qquad + a'_{2, M+1}x_{M+1} + \ldots + a'_{2N}x_N = b'_2$$
$$\vdots$$
$$x_M + a'_{M, M+1}x_{M+1} + \ldots + a'_{MN}x_N = b'_M$$
$$c'_{M+1}x_{M+1} + \ldots + c'_N x_N = z - c'_0$$

where $a'_{ij}$, $b'_i$ and $c'_j$ are the modified values of $a_{ij}$, $b_i$ and $c_j$, and $z$ is a constant. This system is completely equivalent to the original system, and it is called the canonical form for the basis $x_1, x_2, \ldots, x_M$. The coefficients $c'_j$ are called the relative cost coefficients.

Let us see what the reduction to the canonical form has achieved. We have chosen the basis variables as $x_1, x_2, \ldots, x_M$, and therefore the remaining variables are set to zero. Then from the canonical form we automatically know that the solution values are

$$x_i = b'_i \text{ for } i = 1, \ldots, M$$

$$x_i = 0 \text{ for } i = M+1, \ldots, N.$$

Furthermore, the objective function value is $c'_0$. The canonical form for any basis effectively solves the system of equations.

### Example 5.3
The following system of three equations in five variables together with an objective function is converted into the canonical form for the basis $x_1, x_2, x_3$

by the following steps:

$$x_1 + 2x_2 - x_3 + 2x_4 - 5x_5 = 2$$
$$2x_1 + 3x_2 + x_3 - 5x_4 + x_5 = 1$$
$$-x_1 - x_2 + 5x_3 + 2x_4 + 7x_5 = 3$$
$$x_1 + x_2 + x_3 + x_4 + x_5 = z.$$

From the first equation

$$x_1 = 2 - 2x_2 + x_3 - 2x_4 + 5x_5$$

and this can be substituted into the remaining three equations, giving:

$$x_1 + 2x_2 - x_3 + 2x_4 - 5x_5 = 2$$
$$-x_2 + 3x_3 - 9x_4 + 11x_5 = -3$$
$$x_2 + 4x_3 + 4x_4 + 2x_5 = 5$$
$$-x_2 + 2x_3 - x_4 + 6x_5 = z - 2.$$

Now use the second of these equations to express $x_2$ as

$$x_2 = 3 + 3x_3 - 9x_4 + 11x_5$$

and substitute for $x_2$ in the first, third and fourth equations, giving:

$$x_1 + 5x_3 - 16x_4 + 17x_5 = -4$$
$$x_2 - 3x_3 + 9x_4 - 11x_5 = 3$$
$$7x_3 - 5x_4 + 13x_5 = 2$$
$$- x_3 + 8x_4 - 5x_5 = z + 1.$$

Finally, the third equation is used to express $x_3$ as

$$x_3 = \tfrac{2}{7} + \tfrac{5}{7}x_4 - \tfrac{13}{7}x_5$$

and this is substituted in the first, second and fourth equations to give the canonical form

$$x_1 \qquad\quad - 12\tfrac{3}{7}x_4 - 7\tfrac{5}{7}x_5 = -5\tfrac{3}{7}$$
$$x_2 \quad + 6\tfrac{6}{7}x_4 - 5\tfrac{3}{7}x_5 = 3\tfrac{6}{7}$$
$$x_3 - \tfrac{5}{7}x_4 + 1\tfrac{6}{7}x_5 = \tfrac{2}{7}$$
$$- 7\tfrac{2}{7}x_4 - 3\tfrac{1}{7}x_5 = z + 1\tfrac{2}{7}.$$

For the basis $(x_1, x_2, x_3)$ we immediately obtain the solution $(-5\tfrac{3}{7}, 3\tfrac{6}{7}, \tfrac{2}{7})$ and the objective function value is $-1\tfrac{2}{7}$. It may be noted that for this particular example, $x_1, x_2, x_3$ do not form a feasible basis as the canonical form shows that the solution requires $x_1 = -6\tfrac{9}{7}$ and all variables must have non-negative values.

## 5.6 Improving the basis

It will now be assumed that we already have a basic feasible solution and we wish to examine whether it can be improved. It will be presumed for simplicity that the basic feasible solution includes the variables $x_1, x_2, \ldots, x_M$.

(If this is not the case the suffices can readily be altered so that the basic variables take the suffices 1 to $M$.) We wish to enquire whether it is worth introducing one of the non-basic excluded variables $x_{M+1}, ..., x_N$ and getting rid of one of the existing basic variables, thus moving to a new basis. Clearly it is only worth changing the basis if this will lead to a solution with a better objective function value (i.e. an increase in the case of a maximization problem).

The effect of introducing any of the non-basic variables is measured simply by their relative cost coefficients $c'_{M+1}, ..., c'_N$. This is the great advantage of the canonical form. If one of the non-basic variables is to be introduced it will be made positive (because of the non-negative constraints) and therefore the rate of increase of the objective function is measured by the relative cost coefficient. In other words $c'_j$ measures the net change in the objective function for a unit change in $x_j$. Without the canonical representation the effect of introducing a variable would be measured by the combined effects of the gains from the new variable and the gains and losses through the adjustments to the existing basic variables in the system of equations. The canonical form isolates the potential gains by excluding the basic variables from the objective function.

Let us first consider the case in which it is not worth while to introduce any non-basic variable. The basic solution is optimal if the relative cost coefficients are all negative or zero. For in this case none of the excluded variables can be introduced at a positive level to increase the objective function. This condition for optimality can be stated formally as follows. The basis $(x_1, x_2, ..., x_M)$ is an optimal basis if the relative cost coefficients $c'_s$, $s = M+1, ..., N$ of the non-basic variables are all non-positive i.e., if

$$c'_s \leqq 0 \text{ for } s = M+1, ..., N.$$

Then the solution $x_i = b'_i$, $i = 1, ..., M$ is an optimal solution.

However if there are some positive cost coefficients, say $c'_{sj} > 0$ then the objective function can be increased by introducing $x_j$ into the basis. Any variable $x_j$ for which $c'_j > 0$ will increase the objective function value but by convention we will choose the variable $x_j$ which has the most positive cost coefficient, i.e. for which

$$c'_s = \max_{M+1 \leqq j \leqq N} c'_j \quad \text{for } c'_j > 0.$$

Having decided to introduce $x_s$ into the basis we now need to decide which variable to remove. This is done by a simple rule. The magnitude of $x_s$ is chosen as large as possible provided that none of the constraints are violated. As $x_s$ is the only non-basic variable which is allowed to deviate from zero we can express the relationships between $x_s$ and the basic variables as

$$x_i + a'_{is} x_s = b'_i \quad \text{for } i = 1, ..., M.$$

Now consider the effects on these equations of increasing $x_s$ from zero to some positive value. Since $b_i'$ is a fixed quantity, if all $a_{is}'$ are negative $x_s$ may be increased indefinitely. As $x_s$ has a positive cost coefficient the objective function value would go to $+\infty$, and the solution is unbounded. This is as far as we can go in this case. However, there are usually some $a_{is}'$ values which are positive, and then, to maintain feasibility, we require that $x_i \geq 0$, giving

$$b_i' - a_{is}' x_s \geq 0.$$

This provides a set of upper bounds on $x_s$ as

$$x_s \leq \frac{b_i'}{a_{is}'} \quad \text{for } i = 1, \ldots, M.$$

Let the smallest of these upper bounds on $x_s$ occur for $i = r$, say, then the maximum value of $x_s$ is governed by the relation

$$\max x_s = \min_i \frac{b_i'}{a_{is}'} = \frac{b_r'}{a_{rs}'},$$

the minimization extending only over those values of $i$ $(1 \leq i \leq M)$ for which $a_{is}' > 0$. If we let $x_s$ take on its maximum value $b_r'/a_{rs}'$, the value of $x_r$ is reduced to zero, and we will have brought $x_s$ into the basis and excluded $x_r$.

## 5.7 Transformation to the new canonical form

Having selected the new variable we now need to transform the equations to obtain the new canonical form before the procedure can be repeated. This is achieved quite simply. We divide the $r$th equation by $a_{rs}'$ to express $x_s$ in terms of the variables $x_r, x_{M+1}, \ldots, x_N$. Then we use this equation to eliminate $x_s$ from the remaining equations leading to the following system of equations with coefficients marked by asterisks:

$$
\begin{aligned}
x_1 \phantom{+} + a_{1r}^* x_r \phantom{xxxx} + a_{1,M+1}^* x_{M+1} \phantom{xx} + \ldots \phantom{x} +0+ \ldots + a_{1M}^* x_N &= b_1^* \\
x_2^* + a_{2r}^* x_r \phantom{xxxx} + a_{2,M+1}^* x_{M+1} \phantom{xx} + \ldots \phantom{x} +0+ \ldots + a_{2M}^* x_N &= b_2^* \\
\vdots \phantom{xxxxxxxxxxxxxxxx} \\
+ a_{rr}^* x_r \phantom{xxxx} + a_{r,M+1}^* x_{M+1} \phantom{xx} + \ldots \phantom{x} + x_s + \ldots + a_{rN}^* x_N &= b_r^* \\
\vdots \phantom{xxxxxxxxxxxxxxxx} \\
+ a_{Mr}^* x_r + \ldots + x_M \phantom{x} + a_{M,M+1}^* x_{M+1} \phantom{xx} +0+ \ldots + a_{MN}^* x_N &= b_M^* \\
c^* x_r \phantom{xxxx} + c_{M+1}^* x_{M+1} + \ldots \phantom{x} +0+ \ldots + c^* x_N &= z - c_0^*.
\end{aligned}
$$

We can now repeat the whole procedure with this new basis and new canonical form.

## 5.8 Summary of the procedure for a change of basis

The whole simplex procedure for changing the basis and establishing the optimum can be summarized very briefly. When the system of equations has

c

been transformed into canonical form for a given basis, the iterative procedure is as follows:

Examine the $c'_j$ values and pick out the largest $c'_j$, say $c'_s$.

If $c'_s \leqq 0$, the current solution is optimal.

If $c'_s \geqq 0$, examine the $s$th column of coefficients $a'_{is}$ and pick out those $a'_{is} > 0$. If all $a'_{is} \leqq 0$ the optimal solution is unbounded.

If some $a'_{is} > 0$, form the ratio $\dfrac{b'_i}{a'_{is}}$ for $a'_{is} > 0$ and select the smallest, say $\dfrac{b'_r}{a'_{rj}}$ amongst those $a'_{is} > 0$.

Transform to the new canonical form for the basis including $x_s$ and excluding $x_r$ by taking the $r$th equation and using it to isolate $x_s$ and eliminate it from the other equations.

*Example 5.4*

Maximize $2x_1 + x_2$

subject to $\quad x_1 + 2x_2 \leqq 10$

$\qquad\qquad x_1 + \ x_2 \leqq \ 6$

$\qquad\qquad x_1 - \ x_2 \leqq \ 2$

$\qquad\qquad x_1 - 2x_2 \leqq \ 1$

$\qquad\qquad x_1 \geqq 0, x_2 \geqq 0.$

Fig. 5.2 shows the optimal solution at $x_1 = 4$, $x_2 = 2$ and the objective function value is 10.

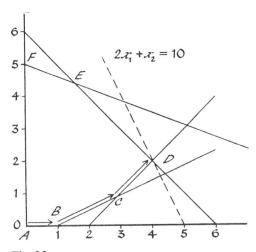

Fig. 5.2

We will now solve this problem by the algebraic procedure of the simplex method. The system is first expressed in standard form by introducing the non-negative slack variables $x_3$, $x_4$, $x_5$ and $x_6$. The introduction of these slack variables also provides a system of equations in canonical form.

$$x_1 + 2x_2 + x_3 \qquad\qquad = 10$$
$$x_1 + x_2 \qquad + x_4 \qquad = 6$$
$$x_1 - x_2 \qquad\quad + x_5 \qquad = 2$$
$$x_1 - 2x_2 \qquad\qquad + x_6 = 1$$

maximize $2x_1 + x_2 = z$.

The slack variables immediately constitute an initial feasible basis, $x_3 = 10$, $x_4 = 6$, $x_5 = 2$, $x_6 = 1$, $x_1 = 0$, $x_2 = 0$. This corresponds to position $A$ in the diagram. The objective function value is zero. The equations are already in canonical form. As $c_1' = 2$ and $c_2' = 1$, the solution is not optimal and we introduce $x_1$ as it has the largest relative cost coefficient. Forming the ratios $b_i'/a_{i1}'$ for $i = 1, \ldots . 4$.

$$\frac{b_1'}{a_{11}'} = 10, \quad \frac{b_2'}{a_{21}'} = 6, \quad \frac{b_3'}{a_{31}'} = 2, \quad \frac{b_4'}{a_{41}'} = 1.$$

The last one is the smallest, so we introduce $x_1$ at value 1 and exclude $x_6$ as the basic variable corresponding to the fourth equation. We now get the new canonical form by using the fourth equation to express $x_1$ as

$$x_1 = 1 + 2x_2 - x_6.$$

The new canonical form is

$$4x_2 + x_3 \qquad - x_6 = 9$$
$$3x_2 \qquad + x_4 \quad - x_6 = 5$$
$$x_2 \qquad + x_5 - x_6 = 1$$
$$x_1 - 2x_2 \qquad + x_6 = 1$$
$$5x_2 \qquad\qquad - 2x_6 = z - 2.$$

The new basis is $x_1 = 1$, $x_3 = 9$, $x_4 = 5$, $x_5 = 1$ and the value of the objective function is 2. This is the point $B$ on the diagram. As $c_2' = 5$, the variable $x_2$ is now introduced, and the minimum positive ratio $\dfrac{b_i'}{a_{i2}'} = 1$ for $i = 3$. Therefore $x_2$ replaces $x_5$. Using the third equation to express $x_2$ we obtain the new system of equations

$$x_3 \qquad -4x_5 + 3x_6 = 5$$
$$x_4 - 3x_5 + 2x_6 = 2$$
$$x_2 \qquad + x_5 - x_6 = 1$$
$$x_1 \qquad + 2x_5 - x_6 = 3$$
$$-5x_5 + 3x_6 = z - 7.$$

The new basis is $x_1 = 3$, $x_2 = 1$, $x_3 = 5$, $x_4 = 2$ and the objective function value is 7. This is the point $C$ on the diagram. The variable $x_6$ is introduced and the variable $x_4$ is removed. Using the second equation to express $x_6$ we obtain the new system

$$
\begin{aligned}
x_3 - \tfrac{3}{2}x_4 + \tfrac{1}{2}x_5 &= 2 \\
+\tfrac{1}{2}x_4 - \tfrac{3}{2}x_5 + x_6 &= 1 \\
x_2 \quad +\tfrac{1}{2}x_4 - \tfrac{1}{2}x_5 &= 2 \\
x_1 \quad +\tfrac{1}{2}x_4 + \tfrac{1}{2}x_5 &= 4 \\
-\tfrac{3}{2}x_4 - \tfrac{1}{2}x_5 &= z - 10.
\end{aligned}
$$

This new basis is $x_1 = 4$, $x_2 = 2$, $x_3 = 2$, $x_6 = 1$. The value of the objective function is 10. We have reached the point $D$, and this solution is optimal.

## 5.9 Obtaining an initial solution

So far we have overlooked the difficulties which may arise in determining an initial basic feasible solution. In the example considered in the previous section, the slack variables provided an immediate initial feasible basis. However if some of the constraints had been 'greater than' conditions with a positive right-hand side, the search for the initial basis cannot be performed by inspection. For instance if the first constraint in Example 5.4 had been

$x_1 + 2x_2 \geqq 10$

the slack variable $x_4$ would have been written as

$x_1 + 2x_2 - x_4 = 10$

and the variable $x_4$ cannot now belong to the initial basis as the solution $x_4 = -10$ violates the non-negative constraints. The same difficulties arise when we have a system of equations to start with rather than inequalities. We could not select a set of variables to constitute a basis, convert to canonical form and guarantee that all the values of the variables will be positive. A special means is needed therefore to obtain an initial feasible solution. This is provided by the introduction of an additional set of $M$ variables called artificial variables, which play no part in the actual problem and will eventually be suppressed.

Suppose that the full system of equations after the introduction of any necessary slack variables is written out in the standard form with $N$ variables and $M$ equations. We now add on an additional set of $M$ artificial variables labelled $x_{N+1}, x_{N+2}, \ldots, x_{N+M}$, one variable for each equation as:

$$
\begin{aligned}
a_{11}x_1 + a_{12}x_2 + \ldots + a_{1N}x_N + x_{N+1} &= b_1 \\
a_{21}x_1 + a_{22}x_2 + \ldots + a_{2N}x_N \quad\quad + x_{N+2} &= b_2 \\
\vdots \\
a_{M1}x_1 + a_{M2}x_2 + \ldots + a_{MN}x_N \quad\quad\quad\quad + x_{N+M} &= b_M.
\end{aligned}
$$

These artificial variables immediately constitute a basis, giving the solution

$$x_{N+i} = b_i, \quad i = 1, 2, ..., M,$$
$$x_j \quad = 0, \quad j = 1, 2, ..., N.$$

Starting with this initial basis we now change the basis to work into the original variables until we have a feasible basis with no artificial variables. This is achieved by the simplex method working with a preliminary objective function which concentrates on eliminating the artificial variables. It does not really matter at this stage which of the original variables are introduced into the objective function and we therefore invent the artificial objective function consisting of minus the sum of the artificial variables and denote it by $w$:

$$-x_{N+1} - x_{N+2} - ... - x_{N+M} = w.$$

This function (sometimes called the infeasibility form) is now to be maximized and it should be observed that the maximum is obtained when all the artificial variables are zero and $w = 0$.

In order to use the simplex method, the system must be converted into canonical form so that the basis variables $x_{N+i}$, $i = 1, ..., M$, do not appear in the objective function. The artificial variables can be expressed in terms of the other variables by the relations

$$x_{N+i} = b_i - \sum_{j=1}^{N} a_{ij} x_j$$

and hence the objective function equation becomes

$$x_1 \sum_{i=1}^{M} a_{i1} + x_2 \sum_{i=1}^{M} a_{i2} + ... + x_N \sum_{i=1}^{M} a_{iN} = w + \sum_{i=1}^{M} b_i.$$

The simplex method can now be applied to minimize $w$. When $w = 0$ all the artificial variables have been eliminated and the basis will consist entirely of original variables. This provides a feasible solution of the original problem and an appropriate canonical form. We could now neglect all the columns of the current canonical form matrix associated with the artificial variables. The original objective function will be written down and converted to its canonical form by substituting for the basis variables where appropriate.

This calculation is often called phase I of the linear programming problem and the solution to the original problem then becomes phase II. The end of phase I provides the start to phase II. If it turns out to be impossible to eliminate all the artificial variables there is no feasible solution to the original problem.

*Example 5.5*

Maximize $x_1 + x_2$

subject to $2x_1 + x_2 \leq 12$
$$x_1 + 2x_2 \geq 2$$
$$x_1, x_2 \geq 0.$$

If slack variables are now introduced this gives the system of equations

$$2x_1 + x_2 + x_3 \quad = 12$$
$$x_1 + 2x_2 \quad -x_4 = 2$$
$$x_1 + x_2 \quad = z.$$

The second equation requires an artificial variable to be introduced, say $x_5$, and we then need to maximize $w = -x_5$ in the system

$$2x_1 + x_2 + x_3 \quad = 12$$
$$x_1 + 2x_2 \quad -x_4 + x_5 = 2$$
$$x_1 + 2x_2 \quad -x_4 \quad = w + 2.$$

The initial basis is $x_3$ and $x_5$ where $x_3 = 12$, $x_5 = 2$ and $w = -2$. The variable $x_2$ is now introduced and $x_5$ is removed from the basis to give the new system of equations:

$$1\tfrac{1}{2}x_1 \quad + x_3 + \tfrac{1}{2}x_4 - \tfrac{1}{2}x_5 = 11$$
$$\tfrac{1}{2}x_1 + x_2 \quad -\tfrac{1}{2}x_4 + \tfrac{1}{2}x_5 = 1$$
$$-x_5 \quad = w.$$

The objective function is now zero and the basis variables are $x_2$ and $x_3$ which do not include any artificial variables. We now remove the column for $x_5$ and return to the original objective function:

$$x_1 + x_2 = z.$$

This is not in canonical form, but the second constraint equation can be used to substitute for $x_2$ so that it is converted to the correct form.

## 5.10 The simplex tableau

In the example of Section 5.8 the results of each iteration were displayed by writing out the transformed equations. However, we do not have to write out the $x_j$ symbols as we merely require their coefficients. The calculations can be laid out compactly in tabular form. The table below shows the quantities relating to a single iteration; a list of basis variables, their values $b'_i$, the objective function value $z$, the coefficients $a'_{ij}$ and the relative cost coefficients. This is called the simplex tableau.

It should be noted carefully that $x_i$, and $a_{ij}, j = 1, \ldots, N$ refers to the variable and coefficients which happen to be in the $i$th row of the table, not the variable $x_i$ as expressed in the original problem.

To move to the next basis we scan the $c'_j$ row to determine the maximum $c'_j$ value, say $c'_s$ for which $c'_s > 0$. Next we look down the $x_s$ column noting $b'_i$ and calculating $b'_i / a'_{is}$ to find the least positive ratio, say $b'_r / a'_{rs}$ ($a'_{rs}$ is called the pivot). This means that $x_s$ comes into the basis and $x_r$ goes out. A new table is written out with values created by the following rules in which the entries for the new table are shown starred. In the basis list $x_s$ replaces $x_r$ in the $r$th row.

In the column of values giving the $b_i$ coefficients and $x_i$ values

$$x_s = \frac{b'_r}{a'_{rs}} = b_s^* \quad \text{(this entry replaces } b_r\text{)}$$

$$x_i = b'_i - a'_{is}\frac{b'_r}{a'_{rs}} = b'_i - a'_{is}b_s^* = b_i^*, \quad i \neq s.$$

The new value of the objective function is given by

$$z = c'_0 + c'_s b_s^*.$$

| Basis | Values | $x_1$ | $x_2$ | ... | $x_s$ | ... | $x_N$ |
|-------|--------|-------|-------|-----|-------|-----|-------|
| $x_1$ | $b'_1$ | $a'_{11}$ | $a''_1$ | ... | $a'_{1s}$ | ... | $a'_{1N}$ |
| $x_2$ | $b'_2$ | $a'_{21}$ | $a'_{22}$ | ... | $a'_{2s}$ | ... | $a'_{2N}$ |
| . | . | . | . | ... | . | ... | . |
| $x_r$ | $b'_r$ | $a'_{r1}$ | $a'_{r2}$ | ... | $\boxed{a'_{rs}}$ | ... | $a'_{rN}$ |
| . | . | . | . | ... | .. | ... | . |
| $x_M$ | $b'_M$ | $a'_{M1}$ | $a'_{M2}$ | ... | $a'_{Ms}$ | ... | $a'_{MN}$ |
| $z$ | $c'_0$ | $c'_1$ | $c'_2$ | ... | $c'_s$ | ... | $c'_N$ |

The coefficients $a_{ij}^*$ of the $i$th row of the table can be expressed in terms of $a'_{ij}$ by the relations

$$a_{rj}^* = \frac{a'_{rj}}{a'_{rs}}$$

$$a_{ij}^* = a'_{ij} - a'_{is}a_{rj}^*, \quad i \neq r,$$

$$c_j^* = c'_j - c_s a_{rj}^*.$$

A new simplex tableau can be written out under the previous tableau and the process repeated until the final solution is obtained.

*Example 5.6*

These formulae may be used to display the tableaux associated with the calculations of Example 5.4.

| Iteration | Basis | Values | $x_1$ | $x_2$ | $x_3$ | $x_4$ | $x_5$ | $x_6$ |
|---|---|---|---|---|---|---|---|---|
| 1 | $x_3$ | 10 | 1 | 2 | 1 | 0 | 0 | 0 |
|  | $x_4$ | 6 | 1 | 1 | 0 | 1 | 0 | 0 |
|  | $x_5$ | 2 | 1 | -1 | 0 | 0 | 1 | 0 |
|  | $x_6$ | 1 | $\boxed{1}$ | -2 | 0 | 0 | 0 | 1 |
|  | $z$ | 0 | 2 | 1 | 0 | 0 | 0 | 0 |
| 2 | $x_3$ | 9 | 0 | 4 | 1 | 0 | 0 | -1 |
| $s=1$ | $x_4$ | 5 | 0 | 3 | 0 | 1 | 0 | -1 |
| $r=6$ | $x_5$ | 1 | 0 | $\boxed{1}$ | 0 | 0 | 1 | -1 |
|  | $x_1$ | 1 | 1 | -2 | 0 | 0 | 0 | 1 |
|  | $z$ | 2 | 0 | 5 | 0 | 0 | 0 | -2 |
| 3 | $x_3$ | 5 | 0 | 0 | 1 | 0 | -4 | 3 |
| $s=2$ | $x_4$ | 2 | 0 | 0 | 0 | 1 | -3 | $\boxed{2}$ |
| $r=5$ | $x_2$ | 1 | 0 | 1 | 0 | 0 | 1 | -1 |
|  | $x_1$ | 3 | 1 | 0 | 0 | 0 | 2 | -1 |
|  | $z$ | 7 | 0 | 0 | 0 | 0 | -5 | 3 |
| 4 | $x_3$ | 2 | 0 | 0 | 1 | $-\frac{3}{2}$ | $\frac{1}{2}$ | 0 |
| $s=6$ | $x_6$ | 1 | 0 | 0 | 0 | $\frac{1}{2}$ | $-\frac{3}{2}$ | 1 |
| $r=4$ | $x_2$ | 2 | 0 | 1 | 0 | $\frac{1}{2}$ | $-\frac{1}{2}$ | 0 |
|  | $x_1$ | 4 | 1 | 0 | 0 | $\frac{1}{2}$ | $\frac{1}{2}$ | 0 |
|  | $z$ | 10 | 0 | 0 | 0 | $-\frac{3}{2}$ | $-\frac{1}{2}$ | 0 |

## *5.11 The revised simplex method

The simplex method which has been presented is not the procedure which is used in computer programs written to solve linear programming problems. There is a modified scheme called the revised simplex method which is computationally more efficient. The revised simplex method exploys exactly the same ideas as the original simplex, but we eliminate some of the calculation involved in the revision of the canonical form.

Suppose we had a very large linear programming problem with, say, 200 variables and 20 constraints. Starting from some initial solution it is very unlikely that all 200 variables will enter the series of bases in moving to the optimum. Yet in the simplex method which has been described we still go on revising the whole of the matrix of coefficients from $a_{ij}$ to $a'_{ij}$ at each iteration. All that we really require at each stage are the relative cost coefficients of the variables not in the basis. In the revised simplex we achieve this reduction calling only on the essential calculations. These include the inverse matrix of the coefficients of the current basis and quantities which we will call the simplex multipliers of the basis. We will define the simplex multipliers first.

The standard system of equations is expressed as:

$$a_{11}x_1 + a_{12}x_2 + \ldots + a_{1N}x_N = b_1$$
$$a_{21}x_1 + a_{22}x_2 + \ldots + a_{2N}x_N = b_2$$
$$\vdots$$
$$a_{M1}x_1 + a_{M2}x_2 + \ldots + a_{MN}x_N = b_M$$
$$c_1x_1 + c_2x_2 + \ldots + c_Nx_N = z - c_0.$$

The canonical form reduced this system of equations to obtain the objective function in the form for the basis $(x_1, x_2, \ldots, x_M)$ as

$$c'_{M+1}x_{M+1} + c'_{M+2}x_{M+2} + \ldots + c'_Nx_N = z - c'_0.$$

However, these relative cost coefficients can be obtained in a different way. Suppose we multiply the first equation by $p_1$, the second equation by $p_2$ and so on multiplying the $M$th equation by $p_M$ and add up the resulting equations we get

$$\left(c_1 + \sum_{i=1}^{M} a_{i1}p_{i1}\right)x_1 + \left(c_2 + \sum_{i=1}^{M} a_{i2}p_{i2}\right)x_2 + \ldots + \left(c_N + \sum_{i=1}^{M} a_{iN}p_{iN}\right)x_N$$
$$= z + \sum_{i=1}^{M} b_ip_i - c_0.$$

Now, in order to make the coefficients of the first $M$ variables zero, we determine the $p_i$ values by solving the system of equations

$$c_j + \sum_{i=1}^{M} a_{ij}p_i = 0 \text{ for } j = 1, \ldots, M$$

(which requires the inverse of the matrix of coefficients $a_{ij}$ for the basis), and we will obtain the relative cost coefficients for the variables excluded from the basis in terms of $p_i$ as

$$c'_j = c_j + \sum_{i=1}^{M} a_{ij}p_i \text{ for } j = M+1, M+2, \ldots, N.$$

The $p_i$ quantities are called the simplex multipliers for the basis $(x_1, x_2, \ldots, x_M)$. It is this connection between the $p_i$ quantities for the basis variables and the $c'_j$ for the excluded variables which is the key to the revised simplex.

We can now outline the details of the revised simplex method. The full derivations will be found in the reference by Garvin. At each iteration we record four sets of information:

the basis, say, $(x_1, x_2, \ldots, x_M)$,
the value of each basis variable, say $(b'_1, b'_2, \ldots, b'_M)$,

the inverse of the basis, with coefficients $v_{ij}$,
the simplex multipliers of the basis, say, $p_1, p_2, \ldots, p_M$.

To move to the next basis we must find the largest relative cost coefficient for the current basis as we did in Section 5.6. This is determined by considering

$$c'_j = c_j + \sum_{i=1}^{M} a_{ij}p_i, \text{ for } j = M+1, M+2, \ldots, M+N.$$

Having found the largest say $c'_s$, we introduce $x_s$ and remove the current variable $x_r$ by finding as before the value $i = r$ for which $b'_i/a'_{is}$ is a minimum considering only values of $a'_{is} > 0$. The values of $a'_{is}$ are determined as

$$a'_{is} = \sum_{k=1}^{M} v_{ik}a_{ks}.$$

We now have the new basis including $x_s$ in the $r$th position in the basis with new values

$$x_i = b^*_i = b^*_i - a'_{is}b^*_r, \quad i \neq r$$

$$x_s = b^*_r = \frac{b'_r}{a'_{rs}}.$$

Finally, in order to up-date the remaining information, the inverse of the new basis with coefficients $v^*_{ik}$ is obtained from the old basis by the relations

$$v^*_{ik} = v_{ik} - \frac{a'_{is}}{a'_{rs}} v_{rk}$$

$$v^*_{rk} = \frac{v_{rk}}{a'_{rs}}.$$

The simplex multipliers for the new basis are also obtained by the straight-forward relationships

$$p^*_i = - \sum_{\substack{k=1 \\ k \neq r}}^{M} v^*_{ki}c_k - v^*_{ri}c_s.$$

(Note that we write $v_{ki}$ rather than $v_{ik}$ as we are dealing with the transposed inverse in this case.)

The procedure is therefore very similar to the method of revision using the full canonical form. But considerable computational time may be saved by using this modified technique. The simplex tableau described in Section 5.10 would need to be adjusted to correspond with this procedure. There are other refinements to the calculations such as the representation of the inverse in a 'product form' which further improve the simplex procedure for computational purposes, but we shall not discuss them here. The details of the best calculating rules can be found in texts devoted solely to linear programming. They all follow the basis to basis improvements enunciated in Sections 5.5 to 5.7.

## *5.12 Difficulties in special cases

The simplex procedure has been presented as if it was bound to work every time. This is not necessarily the case. Some difficulties can arise although they seldom occur in practical problems.

It was assumed that in determining the variable to be removed from the basis the minimum ratio $b'_i/a'_{is}(a'_{is} > 0)$ selected a unique variable. If a tie occurs, say for two variables, then the procedure for reducing the value of one of the variables to zero, reduces the value of the other variable to zero. The other variable is held in the basis at zero value in a suspended state. On the next iteration this variable will be dropped from the basis as its ratio $b'_i/a'_{is}$ is zero, and therefore must be the minimum position ratio. But the elimination of this variable will not increase the objective function value. It is therefore possible in principle to repeat a whole series of these situations and enter a closed cycle of bases, never concluding the solution. This phenomenon is known as degeneracy. We will not study it further here.

We have also ignored any problems which might arise through the linear dependence of the constraints. It has been assumed that the system of equations can always be solved for any basis. Later on in a transportation problem we will show how a constraint equation must be dropped because of the linear dependence. Again linear dependence will not be studied further here. Both these special difficulties are adequately treated in the texts devoted to linear programming theory.

# Appendix

## *5.13 The fundamental theorem of linear programming

The fundamental theorem of linear programming states that the optimal solution to a linear programming problem involves at most $M$ of the $N$ variables in the system of equations, where $M$ is the number of equality constraints when the problem is written in standard form. This will now be proved.

The general linear programming problem in $N$ variables (original as well as slack variables) may be stated as follows:

$$A_1 x_1 + A_2 x_2 + \ldots + A_N x_N = B$$

$$x_1 \geqq 0, \, x_2 \geqq 0, \, \ldots, \, x_N \geqq 0$$

$$c_1 x_1 + c_2 x_2 + \ldots + c_N x_N \quad = z = \text{maximum}$$

where $A_1, A_2, \ldots, A_N$ and $B$ are $M$-dimensional column vectors the elements of which are the coefficients of the $M$ linear equations $(M < N)$. (The constant in the objective function has been dropped for simplicity of notation. Clearly the same problem is being solved.)

We assume that we have an optimal solution in which $r$ variables—say the $r$ first—are positive, the remaining $(N-r)$ variables being zero. If $r \leqq M$, the theorem is automatically true, so let us assume $r > M$. Let

$$(x_1, x_2, \ldots, x_r, x_{r+1}, \ldots, x_N) = (\bar{x}_1, \bar{x}_2, \ldots, \bar{x}_r, 0, \ldots, 0)$$

be this solution; this means that

$$A_1 \bar{x}_1 + A_2 \bar{x}_2 + \ldots + A_r \bar{x}_r = B$$
$$\bar{x}_1 \geqq 0, \bar{x}_2 \geqq 0, \ldots, \bar{x}_r \geqq 0$$
$$c_1 \bar{x}_1 + c_2 \bar{x}_2 + \ldots + c_r \bar{x}_r = \bar{z}.$$

Now a well-known theorem of linear algebra states that a set of vectors is always linearly dependent if their number is greater than their dimension. It follows that there will exist a set of numbers $y_j$, not all zero, such that

$$A_1 y_1 + A_2 y_2 + \ldots + A_r y_r = 0;$$

we may assume that at least one $y_j$ is positive, for if not we could change the sign of all $y_j$ and the equation would still be satisfied. Now let $t = \max (y_j/\bar{x}_j)$, which is clearly a positive number. We can then easily show that

$$A_1 \left( \bar{x}_1 - \frac{y_1}{t} \right) + A_2 \left( \bar{x}_2 - \frac{y_2}{t} \right) + \ldots + A_r \left( \bar{x}_r - \frac{y_r}{t} \right) = B$$

i.e. that the set of numbers

$$\bar{x}_j - \frac{y_j}{t} (j = 1, 2, \ldots, r)$$

is a solution to the original system of equations. The non-negativity of the solution follows from the assumptions

$$\frac{y_j}{\bar{x}_j} \leqq t, \bar{x}_j \geqq 0, \text{ and } t > 0.$$

Moreover, by the definition of $t$, at least one of the numbers is zero, i.e. we have found a feasible solution in which fewer than $r$ variables are positive. The proof is now complete if we can show that the solution is also an optimal one, i.e. that

$$c_1 \left( \bar{x}_1 - \frac{y_1}{t} \right) + c_2 \left( \bar{x}_2 - \frac{y_2}{t} \right) + \ldots + c_r \left( \bar{x}_r - \frac{y_r}{t} \right)$$
$$= c_1 \bar{x}_1 + c_2 \bar{x}_2 + \ldots + c_r \bar{x}_r = \bar{z}$$

which will be true if we can show that

$$c_1 y_1 + c_2 y_2 + \ldots + c_r y_r = 0.$$

Supposing that this were not so we could find a number $u$ such that

$$u(c_1 y_1 + c_2 y_2 + \ldots + c_r y_r) = c_1(uy_1) + c_2(uy_2) + \ldots + c_r(uy_r) > 0$$

or, adding $\sum_{j=1}^{r} c_j \bar{x}_j$ on either side of the inequality sign,

$$c_1(\bar{x}_1 + uy_1) + c_2(\bar{x}_2 + uy_2) + \ldots + c_r(\bar{x}_r + uy_r)$$
$$> c_1 \bar{x}_1 + c_2 \bar{x}_2 + \ldots + c_r \bar{x}_r = \bar{z}.$$

The $(\bar{x}_j + uy_j)$ are easily shown to be a solution to the system of equations for any value of $u$, and by making $|u|$ sufficiently small we could make it a non-negative solution. But this would mean that the $(\bar{x}_j + uy_j)$ would give a larger value of $z$ than the $\bar{x}_j$, contradicting the assumption that the $\bar{x}_j$ are an optimal solution.

We have thus proved that the number of positive variables in an optimal solution can always be reduced so long as $r > M$. Repetition of this reduction must eventually lead to a solution in which at most $M$ variables are positive.

## *5.14 The convex space of feasible solutions

It was also stated that the collection of all feasible solutions constituted a convex set. This will now be proved.

Consider two feasible solutions

$$\bar{x}_1, \bar{x}_2, \ldots, \bar{x}_N$$

and

$$x'_1, x'_2, \ldots, x'_N.$$

We know that

$$\sum_{j=1}^{N} A_j \bar{x}_j = B, \quad \bar{x}_j \geq 0,$$

and

$$\sum_{j=1}^{N} A_j x'_j = B, \quad x'_j \geq 0.$$

If we multiply the first set of equations by $u$ and the second set by $(1-u)$ where $0 \leq u \leq 1$ and add the resulting expressions:

$$\sum_{j=1}^{N} A_j(u\bar{x}_j + (1-u)x'_j) = uB + (1-u)B = B.$$

Therefore all points on the segment between any two feasible solutions is also feasible. As this is the definition of a convex region of points the collection of all feasible solutions constitutes a convex set.

REFERENCES

Garvin, W. W. 1960. *Introduction to Linear Programming*. McGraw-Hill, New York.
Hadley, G. 1962. *Linear Programming*, Addison Wesley, New York.
Wagner, H. M. 1958. The simplex method for beginners. *Op. Res.*, **6**, 364.

## Exercises on Chapter 5

**1** Convert the following problem into standard form:

Minimize $-3x_1-4x_2+2x_3-x_4$

subject to the constraints

$$3x_1+x_2+ x_3 \quad\leqq\quad 7$$
$$4x_1+x_2-6x_3 \quad\geqq\quad 6$$
$$-x_1-x_2+ x_3+x_4 = -4$$

and $x_1 \geqq 0$, $x_2 \geqq 0$.

**2** Solve the following problem:

Maximize $2x_1+x_2$

subject to
$$x_1-2x_2 \leqq 4$$
$$x_1 \leqq 8$$
$$x_1+ x_2 \leqq 15$$
$$-x_1+ x_2 \leqq 6$$
$$x_1 \geqq 0$$
$$x_2 \geqq 0$$

(*a*) graphically

(*b*) by the simplex method.

**3** Obtain an initial feasible basis for the following problem:

Maximize $x_1+2x_2$

subject to
$$x_1+3x_2 \leqq 9$$
$$x_1+ x_2 \leqq 6$$
$$x_1- x_2 \geqq 2$$
$$x_1+ x_2 \geqq 3$$
$$x_1 \geqq 0, x_2 \geqq 0.$$

Display the iterations in tableau form.

**\*4** Consider whether it is possible to use the calculus and Lagrange multipliers on the standard linear programming problem.

# 6 Optimization of non-linear functions

## 6.1 Non-linear optimization techniques

Optimization of non-linear functions is a much more difficult proposition than the linear programming problem. There is no common structure to non-linear functions which can be exploited; non-linearities can take all sorts of forms and there is not a single best technique for all situations. A host of methods have been proposed—each justified in its own context. We will examine six of the most important methods in this chapter.

The general non-linear problem is to find values of the $N$ continuous problem variables $x_1, x_2, ..., x_N$ to minimize a function

$F(x_1, x_2, ..., x_N)$

subject to the $M$ inequality constraints

$G_k(x_1, x_2, ..., x_N) \leqq 0$, for $k = 1, ..., M$,

and the $L$ equality constraints

$H_k(x_1, x_2, ..., x_N) = 0$, for $k = 1, ..., L$,

where the functions $G_k$ and $H_k$ are also non-linear functions. As an illustration of the form of a non-linear optimization problem, the following is a simple example:

Minimize $x_1 x_2 + x_2 e^{x_3}$

subject to $x_1^2 + x_2 \geqq 3$

and $x_1 x_3 = 5$.

The methods of non-linear programming are classified both by the ideas they employ in the search for the optimum and by the class of problem structures which they are able to tackle. The three main search groups consist of gradient methods which use the differential calculus, direct search methods which make no assumption about the analytic nature of the function, and methods which approximate the problem by a linear form so as to use the simplex method of linear programming. We shall first give a simple presentation of gradient methods for solving unconstrained problems and a variety of constrained problems. The methods are called the steepest descent, projected gradient and created response surface techniques, the last being a shrewd way of converting a constrained problem into unconstrained form. These will be followed by a description of an efficient direct search procedure. Finally the methods of separable and approximation programming will be outlined; these employ the simplex method described in the last chapter. All the methods will be concerned with moving from some initial feasible solution to a local optimum. The means of finding an initial feasible point are discussed in Section 6.3.

## 6.2 Local and global optima

The non-linear optimization methods rarely guarantee to find the global optimum. All that can be achieved is to move into a neighbourhood at which the function achieves a local optimum. This is the situation already met in the calculus where the local optimum is characterized by the partial derivatives being zero:

$$\frac{\partial F}{\partial x_i} = 0.$$

It is only when the objective function is convex and the feasible region is also convex that we can be sure that the global optimum will be found. In one dimension a convex objective function will have the form shown in Fig. 6.1. Recalling the definition of convexity, the requirement for the function $f(x)$ to be convex is that it should satisfy the following inequality:

$$f((1-t)y+tz) \leqq (1-t)f(y)+tf(z), \ 0 \leqq t \leqq 1$$

for any two points $x = y$ and $x = z$. This inequality states that the value of the function $f(x)$ at any point between $a$ and $b$, must be less than the value of the function at the point $x$ on the chord joining $(y, f(y))$ and $(z, f(z))$. This is illustrated in Fig. 6.1 where $DE < CE$. The definition may be generalized to a function of several variables. For the objective function $F(X)$ in $N$ variables to be convex, we require that for any two points $(Y), (Z)$

$$Y = (y_1, y_2, ..., y_N)$$
$$Z = (z_1, z_2, ..., z_N)$$

the following inequality must hold:

$$F((1-t)Y+tZ) \leqq (1-t)F(Y)+tF(Z) \text{ for } 0 \leqq t \leqq 1.$$

The definition for the feasible region also being convex follows directly from the definition of convexity. If $X$ is any point on the line joining two points $Y$ and $Z$, $X$ may be expressed as

$$X = tY+(1-t)Z, \quad 0 \leqq t \leqq 1.$$

For the feasible region to be convex we require that if $Y$, $Z$ are feasible then an $X$ as defined above is also feasible and satisfies the constraints expressed by the relations $G_k \leqq 0$ and $H_k = 0$.

When these convexity conditions hold we can start from any feasible point and move steadily 'down-hill' to reach a global minimum; it is impossible to be trapped in the local bowl of a function or 'pinched' into some corner of the feasible region.

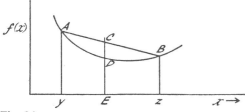

Fig. 6.1

These convexity conditions are restrictive and it is not usually possible to establish that a global optimum has been obtained. It has been suggested that a scatter of initial solutions may be taken and if they always lead to the same local optimum it is likely that they all correspond to the global optimum. But this is expensive in computing time.

## 6.3 Finding an initial feasible solution

Before going on to discuss the methods we will briefly look at how an initial feasible point may be determined. The techniques generally overlook this question assuming that an initial point is available, but this is not always the case.

In principle there is no difficulty about finding an initial feasible point for unconstrained problems. Any set of values for the problem variables should be adequate as the optimization technique will start from this point and work towards the minimum. It is of course wise to choose a point as close to the optimum as possible to reduce computing time. However, when constraints are present and it is not possible to designate a feasible point in advance, an initial starting position must be determined. One safe method is to scan the region of interest over a coarse grid in the problem variables recording for each grid point whether it is feasible and what the objective function value is.

The best feasible grid point can then be used as an initial solution for the optimizing technique. But this approach will be time-consuming if there are a lot of variables.

A more efficient scheme is to use a non-linear optimization method to determine the initial point in much the same way as artificial variables and the simplex method are used in linear programming to obtain a first feasible basis. Consider the case in which the constraints consist of a set of inequalities of the form

$$G_k(X) = G_k(x_1, x_2, ..., x_N) \leq 0, k = 1, ..., M.$$

Suppose we minimize the unconstrained function $W$ where

$$W(X) = \sum_{k=1}^{M} \max (G_k(X), 0).$$

For example, if there were two constraints

$$G_1(x_1, x_2) = x_1^2 - 3x_1x_2 - 7 \leq 0$$

and

$$G_2(x_1, x_2) = x_1 + x_2^2 - 3 \leq 0$$

the function $W(x_1, x_2)$ would be

$$W(x_1, x_2) = \max (x_1^2 - 3x_1x_2 - 7, 0) + \max (x_1 + x_2^2 - 3, 0).$$

Whenever we have a feasible solution $X$ the constraint function values $G_k(X)$ will be zero or negative. If there exists a feasible solution, the minimum value of $W(x_1, x_2)$ is zero. Therefore by finding a point $X$ which minimizes $W(X)$, we are searching for any feasible point. Starting from any initial point we could use an optimization method to move to a solution such that

$$G_k(x_1, x_2, ..., x_N) \leq 0$$

for all $k$, giving the function $W$ a zero value which is its minimum. We would thus determine an initial feasible point.

In a similar way, when equality constraints are present having the form

$$H_k(X) = H_k(x_1, x_2, ..., x_N) = 0, \text{ for } k = 1, ..., L,$$

we can minimize the extended function

$$W(X) = \sum_{k=1}^{M} \max (G_k(X), 0) + \sum_{k=1}^{L} (H_k(X))^2$$

to obtain an initial feasible solution for which $W$ is zero and both sets of constraints will be satisfied. The means of actually minimizing $F'(X)$ will not be described for the moment. It is clearly an unconstrained optimization problem, and the methods for solving these problems have not yet been described. In fact the solution could be determined by the direct search method which is discussed later. But we will assume from now on that we have an initial feasible point and that when one is not known in advance it can be determined in this way.

## 6.4 Unconstrained problems: the failure of the calculus

It is worth pointing out here why the calculus methods can seldom be used satisfactorily on even the unconstrained problems. The calculus states that the optimum will occur at points for which

$$\frac{\partial F}{\partial x_i} = 0.$$

But apart from the difficulty of determining whether such points are local minima, maxima, or saddle points, it is often extremely difficult to solve the derivative equations. Suppose for example we wished to minimize the function

$$F(x_1, x_2) = 2 + 5x_1^2 x_2 + 3e^{-x_1} + 6x_2 e^{-x_1 - x_2}.$$

The partial derivatives provide the equations

$$\frac{\partial F}{\partial x_1} = 10x_1 x_2 - 3e^{-x_1} - 6x_2 e^{-x_1 - x_2} = 0$$

$$\frac{\partial F}{\partial x_2} = 5x_1^2 - 6x_2 e^{-x_1 - x_2} + 6e^{-x_1 - x_2} = 0.$$

But these equations cannot be solved analytically. Numerical procedures have to be used to solve the equations and it is generally a simpler task to use a numerical procedure on the original optimization problem than it is to solve the partial derivative equations.

## 6.5 Steepest descent for unconstrained problems

The unconstrained optimization problem is to find the minimum of some general function $F(x_1, x_2, ..., x_N)$. Our policy will be to start from some trial initial point $X^{(0)}$

$$X^{(0)} = (x_1^{(0)}, x_2^{(0)}, ..., x_N^{(0)})$$

and move steadily downwards one step at a time until we have reached the minimum. We have to decide at each step the direction in which to move and the distance we should move before reassessing whether we are on the correct track. Let the vector $D^{(0)}$

$$D^{(0)} = (d_1^{(0)}, d_2^{(0)}, ..., d_N^{(0)})$$

denote the direction in which to move from the point $X^{(0)}$ and let $l^{(0)}$ be the length of the step. The vector $D^{(0)}$ will be normalized so that its length is unity, *i.e.* $\Sigma d_i^2 = 1$. Then the position reached at the end of the first step is $X^{(1)}$ where

$$X^{(1)} = X^{(0)} + l^{(0)} . D^{(0)}.$$

Similarly we will move to the next point $X^{(2)}$ by determining a new direction $D^{(2)}$ and a length $l^{(2)}$. In general the $(t+1)$th point will be reached from the $t$th point by the relation

$$X^{(t+1)} = X^{(t)} + l^{(t)} . D^{(t)}.$$

The descent follows a zigzag pattern towards a minimum as illustrated in Fig. 6.2.

We now need to determine the components of the unit vector $D^{(t)}$ and $l^{(t)}$. Clearly a sensible policy is to move down in the direction of steepest descent. It is shown in an Appendix (p. 95) that the direction of steepest descent is in

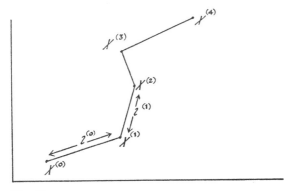

Fig. 6.2

fact given by components proportional to the negative of the partial derivatives evaluated at the point, the constant of proportionality being chosen to ensure the vector has unit length. We therefore set

$$d_i^{(t)} = - \left[ \frac{\partial F}{\partial x_i} \middle/ \left\{ \sum_{k=1}^{N} \left( \frac{\partial F}{\partial x_k} \right)^2 \right\}^{\frac{1}{2}} \right]_{X = X^{(t)}} \quad \text{for } i = 1, ..., N.$$

It should be noted that the direction of steepest ascent points in the opposite direction and the steepest ascent vector has the same formula for its components with the opposite sign. Various methods for determining the quantities $l^{(t)}$ have been proposed. It is sensible to go in the direction $D^{(t)}$ as long as $F(X)$ is being reduced. Here we will adopt a simple and robust search scheme. We set off in the direction $D^{(t)}$ with a trial step length $l$ and evaluate the objective function. If this is reduced we double the distance and re-evaluate the objective function. Then we double the distance again and so on until $F(X)$ increases. If on the first evaluation $F(X)$ increases we halve the distance and then halve it again and so on. The case for the doubling and halving procedure is that it takes care to some extent of any misjudgements about the correct scale of the initial step length. For instance if the step length is too small we will not take a vast number of steps to reach the minimum in the direction $D^{(t)}$.

The computational procedure can now be formalized. We are at point $X^{(t)}$ and we have a proposed trial step length $l$ and a measure of accuracy for the optimum $\delta$. First we check to see if the objective function will be reduced in the direction $D^{(t)}$. If

$$F(X^{(t)}+\delta.D^{(t)})>F(X^{(t)})$$

it shows that no reduction is possible. Since $D^{(t)}$ is the direction of steepest descent, it means that $X^{(t)}$ is the optimum. When this inequality does not hold we evaluate $F(X^{(t)}+l.D^{(t)})$ and compare this with $F(X^{(t)})$. If

$$F(X^{(t)}+lD^{(t)})<F(X^{(t)}),$$

we evaluate $F(X^{(t)}+2l.D^{(t)})$ and compare it with $F(X^{(t)}+l.D^{(t)})$. We continue increasing the step length by a factor of 2 until we have obtained a point $(X^{(t)}+2^m lD^{(t)})$ such that

$$F(X^{(t)}+2^{m+1}lD^{(t)}) \geqq F(X^{(t)}+2^m lD^{(t)})<F(X^{(t)}+2^{m-1}lD^{(t)})$$

and $F(X^{(t)}+2^k lD^{(t)})<F(X^{(t)}+2^{k-1}lD^{(t)})$ for $k = 1, ..., m$. This gives $l^{(t)} = 2^m.l$. Otherwise if $F(X^{(t)}+lD^{(t)})>F(X^{(t)})$ it means we must try smaller steps and try multiplying $l$ by $\frac{1}{2}$, $\frac{1}{4}$, .... Thus we evaluate the sequence $F(X^{(t)}+(l/2^k)D^{(t)})$ for $k = 1, 2, ..., m$ until the following inequalities hold:

$$F(X^{(t)}+(l/2^{m-1})D^{(t)})>F(X^{(t)})$$

and $F(X^{(t)}+(l/2^m)D^{(t)})<F(x^{(t)})$.

This gives

$$l^{(t)} = l/2^m.$$

We thus determine the new point $X^{(t+1)}$ from the point $X^{(t)}$ using $D^{(t)}$ and $l^{(t)}$.

Ultimately it will not be possible to descend any further in any direction when the partial derivatives are all zero. In practice we will halt the procedure when the step length is less than the above measure of tolerance $\delta$.

It is important to choose the quantities $\delta$ and $l$ in a sensible ratio. If $l = 10$, say, and $\delta = 0\cdot001$ a lot of computing time can be wasted cutting down $l$ by factors of $2^m$ near the optimum solution. A ratio of $\delta/l = 1/5$ might be a reasonable choice. Alternatively a dynamic scheme could be used to alter $l$ in the course of the iterations depending on the actual step length taken at the previous iteration. Some possibilities are considered in the exercises.

*Example* 6.1

The following function in two variables is to be minimized

$$F(x_1, x_2) = x_1^2+4x_2^2-4x_1-24x_2+44.$$

We will illustrate one iteration of the method of steepest descent choosing the origin $(0, 0)$ as the initial point and an initial step length of unity. The derivatives are

$$\frac{\partial F}{\partial x_1} = 2x_1 - 4$$

$$\frac{\partial F}{\partial x_2} = 8x_2 - 24$$

and the value of the objective function at the origin is 44. The direction of steepest descent at the origin is (4, 24). To convert this to a unit vector we divide through by its length $\sqrt{(4^2 + 24^2)}$, giving

$$\left(\frac{1}{\sqrt{37}}, \frac{6}{\sqrt{37}}\right) = (0 \cdot 164, 0 \cdot 984)$$

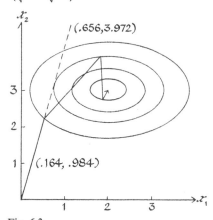

Fig. 6.3

For step length 1, $F(0 \cdot 164, 0 \cdot 984) = 23 \cdot 6 < 44$.
For step length 2, $F(0 \cdot 328, 1 \cdot 968) = 11 \cdot 0 < 23 \cdot 6$.
For step length 4, $F(0 \cdot 656, 3 \cdot 936) = 12 \cdot 19 > 11 \cdot 0$.
Therefore we take the point (0·328, 1·968) as the next point $X^{(1)}$, and repeat the procedure. The direction of steepest descent from this point is (3·34, 8·304), which is converted to a unit vector and the search process is repeated. It will be noted that the direction of movement has changed, and in fact the track that is being followed zigzags across the series of ellipse contours as illustrated in Fig. 6.3.

## *6.6 More advanced descent methods for unconstrained problems

The steepest descent method illustrates the basic idea of descent methods for unconstrained non-linear optimization. But steepest descent is not the best or the only descent method; more sophisticated procedures have been invented for descending over a surface and avoiding the zigzag pattern which can occur in tracing the optimum as happened in the above example. We will briefly mention the nature of these more advanced methods here; the details can be found in the texts specially devoted to non-linear programming.

The steepest descent method takes account of the first derivatives of the function only, but it may be worth while to study the way in which these derivatives change and these changes are embodied in the second derivatives. The first technique to employ second derivatives is known as the Newton-Raphson method. Starting from some current point $X^{(t)}$ we estimate where the minimum lies by using the first three terms in the Taylor series. The minimum is assumed to be at some point $(X^{(t)}+E)$ where

$$(X^{(t)}+E) = (x_1^{(t)}+e_1, x_2^{(t)}+e_2, ..., x_N^{(t)}+e_N).$$

Then

$$F(X^{(t)}+E) = F(X^{(t)}) + \sum_{i=1}^{N} e_i \left[\frac{\partial F}{\partial x_i}\right]_{X=X^{(t)}} + \frac{1}{2} \sum_{i=1}^{N} \sum_{j=1}^{N} e_i e_j \left[\frac{\partial^2 F}{\partial x_i \partial x_j}\right]_{X=X^{(t)}}.$$

We will write this more concisely as

$$F(X^{(t)}+E) = F(X^{(t)}) + \sum_{i=1}^{N} e_i d_i + \frac{1}{2} \sum_{i=1}^{N} \sum_{j=1}^{N} e_i e_j c_{ij}$$

where $d_i$ and $c_{ij}$ have obvious definitions. We wish to find the value of the displacement vector $E$ to minimize the function of $E$, $F(X^{(t)}+E)$. At the minimum we know that the derivatives with respect to $e_i$ will be zero. Therefore assuming $F(X^{(t)}+E)$ is the minimum we differentiate $F(X^{(t)}+E)$ with respect to $e_i$ for $i = 1, ..., N$ to obtain the equations

$$d_i + \sum_{j=1}^{N} c_{ij} e_j = 0, \quad i = 1, 2, ..., N.$$

As these equations are linear in the components $e_j$, their solution can be calculated by inverting the matrix coefficients $c_{ij}$ to obtain

$$e_i = - \sum_{j=1}^{N} (c)_{ij}^{-1} d_j, \quad i = 1, 2, ..., N$$

where $(c)_{ij}^{-1}$ is the $(i, j)$th element in the inverse of the matrix of coefficients $c_{ij}$. The determination of the $e_i$ values enables us to move to the next point $X^{(t+1)}$ as

$$X^{(t+1)} = X^{(t)} + E.$$

The Newton-Raphson method can be very efficient in the right circumstances. But if the initial point $X^{(t)}$ is a poor estimate of the minimum the method often fails to converge. Even when it does converge it may reach a point which is not a minimum as the derivation of an iteration depends only on the fact that the first derivatives of $F(X)$ are zero at the minimum, and the first derivatives are zero at all minima, maxima and saddle points of the objective function.

*Example 6.2*

Suppose we wished to minimize the function of two variables

$$F(x_1, x_2) = 2x_1^2 + x_2^2 + 3x_1x_2 + 10$$

and we are currently at the point $(2, 1)$. Calculating first and second derivatives at this point we get

$$\frac{\partial F}{\partial x_1} = 4x_1 + 3x_2 = 11$$

$$\frac{\partial F}{\partial x_2} = 2x_2 + 3x_1 = 8$$

$$\frac{\partial^2 F}{\partial x_1^2} = 4$$

$$\frac{\partial^2 F}{\partial x_2^2} = 2$$

$$\frac{\partial^2 F}{\partial x_1 \partial x_2} = 3.$$

Now putting these values into the two equations for the displacements $e_1$ and $e_2$ we get

$$11 + 4e_1 + 3e_2 = 0$$
$$8 + 3e_1 + 2e_2 = 0$$

giving

$$e_1 = -2$$
$$e_2 = -1.$$

The new point is obtained by adding $e_1$, $e_2$ to the current point to give the position $(0, 0)$. The derivatives values here are zero and it may appear that we are at a minimum. However, as

$$\frac{\partial^2 F}{\partial x_1^2} \cdot \frac{\partial^2 F}{\partial x_2^2} - \left( \frac{\partial^2 F}{\partial x_1 \partial x_2} \right)^2 = 8 - 9 < 0$$

at the origin, we have in fact reached a saddle point. It should be noted that we could not have reached the origin in a single step of the steepest descent method.

To avoid convergence to a stationary point which is not a minimum, and to ensure that an iteration does not make the value of the objective function worse, the displacement $E$ can be used simply as a direction of search. We then search the function

$$F(X^{(t)} + l.E)$$

in the direction $E$ for varying values of $l$ in exactly the same way as for the steepest descent method.

Further improvements to these descent methods have been developed which employ the ideas of conjugate gradients for quadratic functions and adjust the scales of the variables as the iterations proceed. Perhaps the most successful of these methods originated with Davidon and was fully established by Fletcher and Powell. The method focusses on improving the $(c^{-1})_{ij}$ matrix. These more advanced descent methods are well reviewed in the reference by Powell.

## *6.7 Projected gradient methods for linear constraints

When constraints are present, the steepest descent method may fail because the direction of descent may point outside the feasible region. One means of

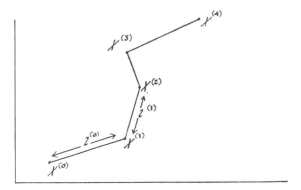

Fig. 6.4

keeping within the feasible region is to steer the direction round inside the constraint boundary. This is the intention of the projected gradient methods. Fig. 6.4 illustrates the requirement, for linear constraints. The descent track $X^{(0)}$, $X^{(1)}$, $X^{(2)}$, $X^{(3)}$, follows the steepest descent until it encounters constraints, then it is necessary to bend the track along the edge where this is advantageous. It will be seen that the method is geometrically quite straightforward as one can proceed in the direction of projected descent along the constraint keeping within the feasible region. The idea of projection can also be applied in the non-linear case, but the scheme is more complex and less successful: it is not considered here.

The basic procedure for the projected gradient method is similar to the steepest descent method except that, if, in searching along the descent direction, a constraint is encountered, we must project the direction along the constraint boundary. We therefore need a formula for determining when we will reach the next constraint and a means of projecting a descent direction.

We wish to minimize the function $F(X)$ subject to a set of linear constraints which will be expressed as the $M$ inequalities

$$a_{11}x_1 + a_{12}x_2 + \ldots + a_{1N}x_N \leqq b_1$$
$$a_{21}x_1 + a_{22}x_2 + \ldots + a_{2N}x_N \leqq b_2$$
$$\vdots$$
$$a_{M1}x_1 + a_{M2}x_2 + \ldots + a_{MN}x_N \leqq b_M.$$

These constraints may also be written in matrix form as

$$A.X \leqq B.$$

Suppose we start at point $X^{(0)}$ interior to the constraint region and wish to move in the direction of steepest descent $D^{(0)}$. Then the distance $L^{(0)}$ which we can move before reaching a constraint is governed by the relation

$$A(X^{(0)} + L^{(0)}D^{(0)}) \leqq B$$

i.e.

$$\sum_{j=1}^{N} a_{ij}(x_j^{(0)} + L^{(0)}d_j^{(0)}) \leqq b_i \text{ for all } 1 \leqq i \leqq M.$$

Therefore

$$L^{(0)} \leqq \frac{b_i - \sum\limits_{j=1}^{N} a_{ij}x_j^{(0)}}{\sum\limits_{j=1}^{N} a_{ij}d_j^{(0)}} \text{ for } 1 \leqq i \leqq M.$$

Hence

$$L^{(0)} = \min_{1 \leqq i \leqq M} \left[ \frac{b_i - \sum\limits_{j=1}^{N} a_{ij}x_j^{(0)}}{\sum\limits_{j=1}^{N} a_{ij}d_j^{(0)}} \right].$$

We can now search for the best point along the line $X^{(0)}$ to $X^{(0)} + L^{(0)}D^{(0)}$ by taking increasing steps of length $l$, $2l$, $4l$, ..., as we did for the steepest descent method. If we proceed up to the constraint, then the point $X$, $X^{(1)}$ will be

$$X^{(1)} = X^{(0)} + L^{(0)}D^{(0)}$$

which lies on a constraint boundary and we will need to perform a projection at the next iteration.

In general suppose the point $X^{(t)}$ lies on a constraint boundary. We need to determine a new descent direction and if necessary project the direction onto a constraint surface. Let the direction of steepest descent be

$$D^{(t)} = (d_1^{(t)}, d_2^{(t)}, \ldots, d_N^{(t)})$$

where

$$d_i^{(t)} = \frac{-\partial F/\partial x_i}{\left\{ \sum\limits_{i=1}^{N} (\partial F/\partial x_i)^2 \right\}^{\frac{1}{2}}}.$$

Then it is possible to identify which constraints the point $X^{(t)}$ lies on by examining for which rows of $A$ the equality

$$A.X^{(t)} = B$$

holds. Suppose there are $r$ such binding equalities; we then form a matrix $M$ of dimensions $N \times r$ whose columns are formed by the elements of the co-efficients of the rows in $A$ corresponding to the binding constraints. The steepest descent direction $D^{(t)}$ along the constraints is obtained by multiplying $D$ by the $N$-dimensional square matrix $P^{(t)}$ where

$$P^{(t)} = I - M(M'.M)^{-1}M'.$$

$I$ being the identity matrix with unit entries in the diagonal and zeros otherwise. The construction of this rather complicated projection matrix is justified in an Appendix (p. 96). The projected direction is therefore the vector

$$P^{(t)}.D^{(t)}$$

(where $D^{(t)}$ is written as a column vector representing the steepest descent direction). When $P^{(t)}.D^{(t)}$ has been converted into a unit vector we then search in this direction in the usual way. If all the elements of $P^{(t)}.D^{(t)}$ are zero, we are either at the optimum or one of the constraints should be dropped from $M$. These alternatives are distinguished by examining the $r$ elements of the vector

$$(M'.M)^{-1}M'D^{(t)}$$

which are the components of the vector representing the difference between $D^{(t)}$ and $P^{(t)}.D^{(t)}$ expressed in terms of the outward normals to the binding hyper-plane constraints. If all these are positive the minimum has been found. If any element is negative, this component is pointing in towards the feasible region and the corresponding row of $A$ should be eliminated from $M$ and the procedure repeated. The details of these formulae are explained fully in the Appendix (p. 96).

When the steepest descent direction has been projected we determine the furthest distance we can move by a formula for $L^{(t)}$ similar to the formula for $L^{(0)}$ neglecting constraints on which the point $X^{(t)}$ lies. We then search in this direction up to the limit $L^{(t)}$ by the standard procedure.

*Example* 6.3

Maximize $F(x_1, x_2) = (x_1 - 3)^2(x_2 - 4)$

subject to $x_1 \quad\ \leqq 2$

$\qquad x_1 + x_2 \leqq 3$

$\qquad x_2 \quad\ \leqq 2$

taking $X^{(0)} = (0.2, 1.8)$ as the starting point.

$$\frac{\partial F}{\partial x_1} = 2(x_1-3)(x_2-4)$$

$$\frac{\partial F}{\partial x_2} = (x_1-3)^2.$$

(Note that we will take the positive value of the partial derivatives as we are maximizing and therefore want the direction of steepest ascent.) The constraints may be expressed as $A.X \leq B$, which are written out in full as:

$$\begin{pmatrix} 0 & 1 \\ 1 & 1 \\ 1 & 0 \end{pmatrix} \begin{pmatrix} x_1 \\ x_2 \end{pmatrix} \leq \begin{pmatrix} 2 \\ 3 \\ 2 \end{pmatrix}.$$

*First iteration*
The initial point satisfies all the constraints so we set off in the direction of steepest ascent. The direction of steepest ascent is

$D^{(0)} = (0.844, 0.54).$

The maximum distance we can move in this direction is given by the formula

$L^{(0)} = \min (0.371, 0.723, 2.24) = 0.371.$

For example the first number $0.371$ is calculated by assessing when we would reach the first constraint in the matrix $x_2 \leq 2$. As we start from the point $(0.2, 1.8)$ the coordinates of the points a distance $l$ from $(0.2, 1.8)$ in the direction of steepest descent are

$(x_1, x_2) = (0.2, 1.8)+l(0.844, 0.54).$

Thus we require

$x_2 = 1.8+0.54l \leq 2$

i.e. $l \leq 0.371$.

We will assume (correctly in this problem) that we always move as far as the first binding constraint and we omit in this iteration and in the later iterations the search along the direction before we reach the constraint. Therefore

$X^{(1)} = (0.2, 1.8)+0.371 (0.844, 0.54) = (0.512, 2.0),$

which is on the boundary of the first constraints in the matrix $AX \leq B$.

*Second iteration*
The descent vector is $D^{(1)} = (9.95, 6.18).1/[(9.95)^2+(6.18)^2]^{\frac{1}{2}}$. The projection matrix $P^{(1)}$ is calculated using the first row of $A$ as the column for $M$,

$$P^{(1)} = \begin{pmatrix} 1 & 0 \\ 0 & 1 \end{pmatrix} - \begin{pmatrix} 0 \\ 1 \end{pmatrix}\left[ (0, 1)\begin{pmatrix} 0 \\ 1 \end{pmatrix}\right]^{-1}(0, 1)$$

$$= \begin{pmatrix} 1 & 0 \\ 0 & 1 \end{pmatrix} - \begin{pmatrix} 0 \\ 1 \end{pmatrix}1.(0, 1) = \begin{pmatrix} 1 & 0 \\ 0 & 1 \end{pmatrix} - \begin{pmatrix} 0 & 0 \\ 0 & 1 \end{pmatrix} = \begin{pmatrix} 1 & 0 \\ 0 & 0 \end{pmatrix}$$

$P^{(1)}D^{(1)} = (0.95, 0).1/[(9.95)^2+(6.18)^2]^{\frac{1}{2}}.$

When this is converted to a unit vector, it is the direction (1, 0).

The new maximum distance which we can move in this direction is

$$L^{(1)} = \min\,(0{\cdot}488,\ 1{\cdot}488) = 0{\cdot}488.$$

(As we are moving along the first constraint we do not consider it in the assessment of minimum distance.)

Hence $X^{(2)} = (0{\cdot}512,\ 2{\cdot}00) + 0{\cdot}488\,(1,\ 0) = (1,\ 2)$. This lies on the boundary of both constraints 1 and 2.

*Third iteration*

The descent direction is $D^{(2)} = (0{\cdot}895,\ 0{\cdot}446)$. We must now use both the first and second rows of $A$ to construct $P^{(2)}$.

$$P^{(2)} = \begin{pmatrix} 1 & 0 \\ 0 & 1 \end{pmatrix} - \begin{pmatrix} 0 & 1 \\ 1 & 1 \end{pmatrix} \left[ \begin{pmatrix} 0 & 1 \\ 1 & 1 \end{pmatrix}\begin{pmatrix} 0 & 1 \\ 1 & 1 \end{pmatrix} \right]^{-1} \begin{pmatrix} 0 & 1 \\ 1 & 1 \end{pmatrix} = \begin{pmatrix} 0 & 0 \\ 0 & 0 \end{pmatrix}$$

giving $(M'M)^{-1}M'D^{(2)} = (-0{\cdot}446,\ 0{\cdot}865)$.

Therefore we are not at the optimum and the sign of the first component indicates that the first row of $A$ is no longer binding. To keep the superscripts in accordance with the iteration number we write

$$X^{(3)} = X^{(2)}.$$

*Fourth iteration*

The descent direction is the same as before

$$D^{(3)} = D^{(2)}.$$

The projection matrix is constructed using only the second constraint as

$$P^{(3)} = \begin{pmatrix} \frac{1}{2} & -\frac{1}{2} \\ -\frac{1}{2} & \frac{1}{2} \end{pmatrix}.$$

$$P^{(3)}.D^{(3)} = (0{\cdot}223,\ -0{\cdot}223)$$

which, expressed as a unit vector is $(0{\cdot}707,\ -0{\cdot}707)$.

The greatest distance we can move is

$$L^{(3)} = 1{\cdot}41$$

$$X^{(4)} = (1,\ 2) + 1{\cdot}41\,(0{\cdot}707,\ -0{\cdot}707) = (2,\ 1)$$

which is at the intersection of the second and third constraints.

*Fifth iteration*

The descent direction is $D^{(4)} = (6/\sqrt{37},\ 1/\sqrt{37})$.

The projection matrix $P^{(4)} = \begin{pmatrix} 0 & 0 \\ 0 & 0 \end{pmatrix}$, and

$$(M'M)^{-1}M'D^{(4)} = \begin{pmatrix} 0 & 1 \\ 0 & -1 \end{pmatrix} \cdot \begin{pmatrix} 6/\sqrt{37} \\ 1/\sqrt{37} \end{pmatrix} = \begin{pmatrix} 1/\sqrt{37} \\ 5/\sqrt{37} \end{pmatrix}$$

which has no negative component. (2, 1) is therefore the maximum. The sequence of iterations are illustrated geometrically in Fig. 6.5.

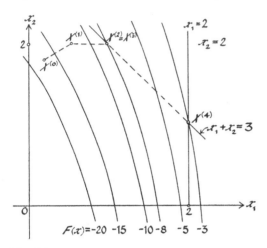

Fig. 6.5

## 6.8 Created response surface techniques for general constraints

The created response surface is a shrewd scheme for overcoming the difficulties of dealing with non-linear constraints. The proposal is to build the constraints into the objective function as penalty functions so that when the constraints would be violated a heavy cost is incurred in the objective function. We are thus steered away from constraints by operating simply on the revised objective function without any explicit restrictions. This method has been developed by Fiacco and McCormick and it is particularly appropriate when there are several non-linearities in the constraint functions.

First let us consider the situation with inequality constraints. Suppose we wish to minimize the function

$$F(X) = F(x_1, x_2, ..., x_N)$$

subject to the constraints

$$G_k(X) = G_k(x_1, x_2, ..., x_N) \leqq 0.$$

Now, let us create the new objective function (or response surface), $P(X, r)$ as

$$P(X, r) = F(X) - r \sum_{k=1}^{M} \frac{1}{G_k(X)}$$

where $r$ is some positive number which we will call the constraint weighting factor. This has the effect of introducing a positive penalty cost as we approach the boundary of the feasible region. As the $G_k(X)$ are negative in the feasible region the contribution to $P(X, r)$ from the additional terms

$$-r \sum_{k=1}^{M} \frac{1}{G_k(X)}$$

is a positive quantity at any feasible point. If $r$ is quite small and we are at an interior point, it will be a comparatively small contribution. However, as we approach a constraint boundary, say $G_k(X) = 0$, the magnitude of $1/G_k(X)$ becomes very large. Therefore if we attempt to minimize $P(X, r)$ starting from some initial feasible point we will tend to be turned back from the edge of the constraints and tend not to cross the constraint boundaries. Now it may be that the minimum lies on the constraint boundaries. However, we will still end up at the optimum by working with the created response surface if we solve a series of minimization problems in which the quantity $r$ is judiciously varied. There is a theorem which shows that, if we minimize $P(X, r)$ for an appropriate decreasing sequence of values of $r$, we will obtain a sequence of values of $X$ which come arbitrarily close to the minimum of $F(X)$ as $r$ approaches zero. This theorem is proved in an Appendix (p. 97). Thus one converges to the actual minimum of $F(X)$ on a path of minima of $P(X, r)$.

We begin by choosing an initial point $X^{(0)}$ in the feasible region and choose an initial value for the weighting factor, say $r^{(0)}$. We then use the method of steepest descent to find the minimum of $P(X^{(0)}, r^{(0)})$ which leads to a new value of $X^{(1)}$. Using $X^{(1)}$ as an initial point a new value of $r$, say $r^{(1)} < r^{(0)}$ (for example $r^{(1)} = r^{(0)}/10$) is taken and the method of steepest descent is again used to minimize $P(X^{(1)}, r^{(1)})$. We now choose a new value $r^{(2)} < r^{(1)}$ and determine $X^{(2)}$. The procedure continues in this way, until insignificant changes are occurring in $P$ and $F$ for continuing decreases in the value of $r$. The method will now be illustrated on an elementary example to clarify the procedure.

*Example 6.4*

The function $f(x) = -x$ is to be minimized subject to the constraint $0 \le x \le 1$. This constraint can be written as two 'less than' constraints as

$$x - 1 \le 0$$

and $-x \le 0$.

The created response surface is then

$$P(x, r) = -x + \frac{r}{1-x} + \frac{r}{x}$$

$$= -x + \frac{r}{x(1-x)}.$$

We start by setting $r = 1$ and determine a minimum at $\bar{x}^{(1)}$. We then set $r = 0 \cdot 1$ and determine a minimum at $\bar{x}^{(2)}$ and so on. The form of the curves corresponding to these various values of $r$ are illustrated in Fig. 6.6. The points $\bar{x}^{(1)}$, $\bar{x}^{(2)}$, $\bar{x}^{(3)}$, $\bar{x}^{(4)}$ form the local optima.

It will be seen how each successive $\bar{x}^{(t)}$ value may be used as a starting point for the next iteration. Also, once a pattern of $\bar{x}^{(t)}$ values has been established it will be possible to extrapolate to improve the starting points.

There is no definite rule for deciding the choice of the $r^{(t)}$ quantities. Experience has suggested that the amount of work required to locate the

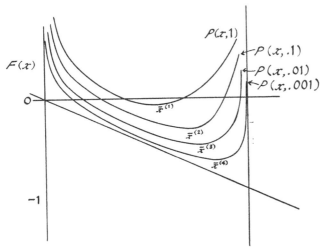

Fig. 6.6

optimum is relatively independent of the rate of decrease of $r$. The faster the reduction, the fewer are the surfaces which have to be optimized, but each minimum is more difficult to locate. Slower reduction necessitates more surfaces each of which is easier to optimize. A further example with numerical results is now illustrated.

*Example 6.5*

Minimize $F(x) = x_1^3 - 6x_1^2 + 11x_1 + x_3$

subject to
$$x_1^2 + x_2^2 - x_3^2 \leqq 0$$
$$-x_1^2 - x_2^2 - x_3^2 + 4 \leqq 0$$
$$x_3 - 5 \leqq 0$$
$$x_1 \geqq 0$$
$$x_2 \geqq 0$$
$$x_3 \geqq 0$$

(Note that the last three constraints may be written as $-x_i \leq 0$ for $i = 1, 2, 3$, to conform to the convention for writing the constraints as 'less than' inequalities.) The global solution to this problem is at $(0, \sqrt{2}, \sqrt{2})$ with the first two constraints and $x_1 \geq 0$ binding. The function $P(x, r)$ is written as

$$P(x, r) = x_1^3 - 6x_1^2 + 11x_1 + x_3 - \frac{r}{x_1^2 + x_2^2 - x_3^2} + \frac{r}{x_1^2 + x_2^2 + x_3^2 - 4}$$

$$- \frac{r}{x_3 - 5} + \frac{r}{x_1} + \frac{r}{x_2} + \frac{r}{x_3}.$$

The following table describes the results of the successive iterations where $r$ is reduced by 10, $10^3$ and $10^6$ respectively.

**Table 6.1**

| Iteration $k$ | $r^{(k)}$ | $x_1^{(k)}$ | $x_2^{(k)}$ | $x_3^{(k)}$ | $F(X^{(k)})$ |
|---|---|---|---|---|---|
| 1 | 1·0 | 0·37896 | 1·68076 | 2·34720 | 5·70854 |
| 2 | 0·1 | 0·10088 | 1·41984 | 1·68317 | 2·73285 |
| 3 | 0·0001 | 0·00302 | 1·41421 | 1·42266 | 1·45581 |
| 4 | $10^{-9}$ | 0·00001 | 1·41421 | 1·41424 | 1·41434 |

The response surface technique can be extended to problems with equality constraints. Suppose we wish to minimize the function $F(X)$ subject to

$$G_k(X) \leq 0 \quad \text{for } k = 1, \ldots, M$$

and

$$H_k(X) = 0 \quad \text{for } k = 1, \ldots, L.$$

We now add a term to the created response surface corresponding to each equality constraint which becomes small on the constraint and large away from it and is controlled by the parameter $r$. A function which satifies these requirements is obtained by squaring $H_k(X)$ and dividing by $\sqrt{r}$. The new function $P(X, r)$ is

$$P(X, r) = F(X) - r \sum_{k=1}^{M} \frac{1}{G_k(X)} + \frac{1}{\sqrt{r}} \sum_{k=1}^{L} (H_k(X))^2.$$

The effect of the multiplier $1/\sqrt{r}$ is to make the equality constraint less dominant than the inequality constraint in the initial stages of the optimization process. Ultimately as $r$ gets smaller we will have to keep nearer and nearer to the equality constraint to ensure that $H_k(X)$ is a small number. It is desirable to square the function $H_k(X)$ as otherwise it may be worthwhile to deviate a long way from the equality constraint, making the component negative and artificially reducing the objective function value.

D

## 6.9 Direct search procedures

The idea of direct search procedures is to work directly with the objective function and avoid the difficulties of calculating gradients of functions. They are much simpler to use than the methods which require partial derivatives and can readily cope with constraints. Indeed they extend naturally to functions that have no analytic or well-defined form. The main drawback is that if there are a lot of variables the computing time required may become large.

The basic scheme is to evaluate the objective function at a series of points which form a carefully directed search across the feasible region. We will assume that we wish to minimize a function of $N$ variables $F(X)$ and initially we will consider the problem without constraints. The search procedure is best described in terms of base points and temporary positions. We start at some initial feasible point which we call the first base point and denote it by

$$X = B^{(0)} = (b_1^{(0)}, b_2^{(0)}, ..., b_N^{(0)}).$$

A step length $\delta_i$ is chosen for each variable $x_i$; in order to use vector notation this will be expressed in the vector $D_i$ whose $i$th component is $\delta_i$, all the rest being zero. We now perturb or vary each variable in turn by amounts $+\delta_i$ or $-\delta_i$ each time accepting the adjustment if it leads to an improvement. When each variable has been perturbed we reach the new base point $B^{(1)}$. The local perturbations are examined first.

We first perturb the variable $x_1$. The objective function is evaluated at $B^{(0)}$ and also at $B^{(0)}+D_1$. If

$$F(B^{(0)}+D_1)<F(B^{(0)})$$

then the point $(B^{(0)}+D_1)$ is called the temporary position and denoted by $T_1^{(0)}$. Otherwise if

$$F(B^0+D_1) \geqq F(B^{(0)})$$

we evaluate $F(B^{(0)}-D_1)$ and if this is less than $F(B^{(0)})$ this is the temporary position. If this also offers no improvement $B^{(0)}$ is designated the temporary position. Thus $T_1^{(0)}$ is determined by one of the three relations:

$$T_1^{(0)} = \begin{cases} B^{(0)}+D_1, & \text{if } F(B^{(0)}+D_1)<F(B^{(0)}) \\ B^{(0)}-D_1, & \text{if } F(B^{(0)}-D_1)<F(B^{(0)})<F(B^{(0)}+D_1) \\ B^{(0)}, & \text{if } F(B^{(0)}) \quad <\min\left(F(B^{(0)}+D_1), F(B^{(0)}-D_1)\right) \end{cases}$$

where the alternative formulae are taken in turn.

Now the next variable $x_2$ is perturbed about the temporary position $T_1^{(0)}$ instead of the original base $B^{(0)}$, and $T_2^{(0)}$ is calculated as the new temporary position. In general the $j$th temporary position $T_j^{(0)}$ is obtained from $T_{j-1}^{(0)}$ by the formulae

$$T_j^{(0)} = \begin{cases} T_{j-1}^{(0)}+D_j, & \text{if } F(T_{j-1}^{(0)}+D_j)<F(T_{j-1}^{(0)}) \\ T_{j-1}^{(0)}-D_j, & \text{if } F(T_{j-1}^{(0)}-D_j)<F(T_{j-1}^{(0)})<F(T_{j-1}^{(0)}+D_j) \\ T_{j-1}^{(0)}, & \text{if } F(T_{j-1}^{(0)}) \quad <\min \left(F(T_{j-1}^{(0)}+D_j), F(T_{j-1}^{(0)}-D_j)\right). \end{cases}$$

This expression covers all $j$, $0 \leq j \leq N$, if we adopt the convention that

$$T_0^{(0)} = B^{(0)}.$$

When all the variables have been perturbed, the last temporary point $T_N^{(0)}$ is designated the second base point $B^{(1)}$,

$$B^{(1)} = T_N^{(0)}.$$

All these exploratory moves which determine the movement from $B^{(0)}$ to $B^{(1)}$ establish a pattern of movement. Now instead of exploring around $B^{(1)}$

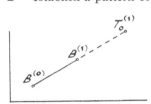

Fig. 6.7

in a similar fashion we assume that the pattern may persist and start the next temporary search to a position not at $B^{(1)}$ but at a point $2(B^{(1)}-B^{(0)})$ from $B^{(0)}$. Thus

$$T_0^{(1)} = B^{(0)}+2(B^{(1)}-B^{(0)}) = 2B^{(1)}-B^{(0)}.$$

This is illustrated in Fig. 6.7.

A local exploration is now carried out around $T_0^{(1)}$, and the equations for determining $T_j^{(1)}$ for $j = 1, \ldots, N$ are the same as the equations for $T_j^{(0)}$ with the superscript 1 replacing zero. Then if the final temporary position $T_N^{(1)}$ is an improvement on the objective function value at $B^{(1)}$ this is established as the new base, i.e.

$$B^{(2)} = T_N^{(1)}, \text{ if } F(T_N^{(1)})<F(B^{(1)}).$$

Assuming this condition holds we now make a further 'double-step' from $B^{(1)}$ beyond $B^{(2)}$ to the temporary position $T_0^{(2)}$ where

$$T_0^{(2)} = 2B^{(2)}-B^{(1)}$$

and perform new exploratory moves around $T_0^{(2)}$.

However if the double jump was a false move and it turns out that the the objective function has increased; i.e.

$$F(T_N^{(1)}) \geq F(B^{(1)})$$

we retreat to the previous base point by setting

$$B^{(2)} = B^{(1)}.$$

The pattern of movement is thus destroyed and the whole procedure is started again treating $B^{(2)}$ as an initial point.

It is clear from these details how the general procedure should be expressed with superscripts $t$ and $(t+1)$ instead of 1 and 2 for moving from $B^{(t)}$ to $B^{(t+1)}$ at iteration $t$. The scheme of extending the pattern ahead to new temporary points enables the step length to be automatically adjusted. When no improvement can be made around some initial point the scale of the step length $\delta_i$ can be halved and the whole procedure repeated until the required accuracy is obtained.

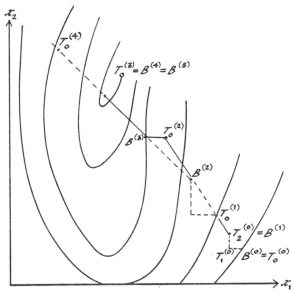

Fig. 6.8

Fig. 6.8 displays how the direct search method would work on an objective function in two variables where the position of the minimum is indicated by the contours. Starting at the base point $B^{(0)}$ we decrease $x_1$ by $\delta_1$ and increase $x_2$ by $\delta_2$ to reach $T_2^{(0)}$ and the new base $B^{(1)}$. We now jump to $T_0^{(1)}$ and start exploring around $T_0^{(1)}$ and thus determine $B^{(2)}$ as it is an improvement on $B^{(1)}$. Next we explore around $T_0^{(2)}$ which is double the distance $B^{(1)}$ to $B^{(2)}$ from $B^{(1)}$. (It may be noted that this jump step length is double the initial jump step length and if we continued in the same direction it would triple, quadruple, etc.) We now explore around $T_0^{(2)}$ to find that only a decrease in $x_1$ is worth while so that we establish $B^{(3)}$. Exploration is now conducted around $T_0^{(3)}$ without any improvement due to the exploration, but as $F(T_2^{(3)}) < F(B^{(2)})$ the new base is $B^{(4)}$ and the temporary position for exploration is $T_0^{(4)}$. At this point we have overshot the minimum, but as exploration around $T_0^{(3)}$ will lead to the result $F(T_2^{(4)}) > F(B^{(3)})$, we are returned to the old base as $B^{(5)} = B^{(4)}$. We would now search around $B^{(5)}$ possibly with a smaller step length.

The layout for numerical calculations is illustrated in the following worked example.

*Example 6.6*

Minimize the function $F(x_1, x_2)$ from Example 6.1 by the direct search method.

$$F(x_1, x_2) = x_1^2 + 4x_2^2 - 4x_1 - 24x_2 + 44.$$

Choose $\delta_1 = \delta_2 = 1$, and let $(0, 0)$ be the initial base point.

Evaluate base $B^{(0)} = T_0^{(0)}$, $F(0, 0) = 44$.

---

Explore round base $B^{(0)} = T_0^{(0)}$,  $F(1, 0) = 41$  accept

$F(1, 1) = 21$  accept

New base $B^{(1)} = (1, 1)$

Temporary position $T_0^{(1)} = (2, 2)$ with $F(2, 2) = 8$.

---

Explore round $T_0^{(1)}$,  $F(3, 2) = 9$  reject

$F(1, 2) = 9$  reject

$F(2, 3) = 4$  accept

New base $B^{(2)} = (2, 3)$

Temporary position $T_0^{(2)} = (4, 6) - (1, 1) = (3, 5)$ with $F(3, 5) = 21$.

---

Explore round $T_0^{(2)}$,  $F(4, 5) = 24$  reject

$F(2, 5) = 20$  accept

$F(2, 6) = 40$  reject

$F(2, 4) = 8$  accept

New base $B^{(3)} = (2, 3) = B^{(2)}$ as $F(2, 4) > F(B^{(2)})$.

---

Explore round base $B^{(3)} = T_0^{(3)}$, $F(3, 3) = 5$  reject

$F(1, 3) = 5$  reject

$F(2, 4) = 8$  reject

$F(2, 2) = 8$  reject

Hence optimum is at the point $(2, 3)$.

The direct search method provides a simple means of handling constraints.

The actual value of $F(X)$ is replaced by a very large value wherever the constraints are not satisfied. This pseudo-value must be proportional to the extent to which the constraints are violated so that the procedure will force the search back into the region where the inequalities can be obeyed.

## *6.10 Separable programming

The method of separable programming provides a means of using the simplex method of linear programming on a non-linear problem. It is only suitable for a class of functions known as separable functions which consist of a sum

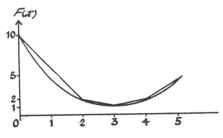

Fig. 6.9

of functions of single variables. The objective and constraint functions must have the form

$$F(x_1, x_2, ..., x_N) = \sum_{i=1}^{N} f_i(x_i)$$

and

$$G_k(x_1, x_2, ..., x_N) = \sum_{i=1}^{N} g_{ki}(x_i) \leqq 0$$

where the functions $f_i$ and $g_{ki}$ are independent of $x_1, x_2, ..., x_{i-1}, x_{i+1}, ..., x_N$.
For example, the function

$$F(x_1, x_2) = x_1^2 + e^{x_1} + 3x_2^3$$

is separable, but the function

$$F(x_1, x_2) = 5x_1 + 2x_2 + x_1^2 x_2$$

is not. It will also be assumed that the $x_i$ values must lie in a certain range so that

$$b_i \leqq x_i \leqq B_i.$$

The first step towards translating the problem into a linear framework is to approximate each function $f_i(x_i)$ and $g_{ki}(x_i)$ by a series of linear segments. The x-axis interval $(b_i, B_i)$, is divided into points $x_i^{(1)}, x_i^{(2)}, ..., x_i^{r(i))}$ as illustrated in Fig. 6.9, where $x_i^{(1)} = b_i$ and $x_i^{r(i)} = B_i$. The function is approximated as a set of straight lines joining the $f_i(x_i)$ values corresponding to the points.

We require enough points between $b_i$ and $B_i$ to ensure reasonable correspondence between the lines and the original curve. Then the 'straight line' function $f_i(x_i)$ may be expressed for any value of $x_i$ as a linear combination of the two points say $x_i^{(j)}$, $x_i^{(j+1)}$ which lie on either side of $x_i$. By simple linear interpolation $f_i(x_i)$ is approximated as $\hat{f}_i(x_i)$, where

$$\hat{f}_i(x_i) = f_i(x_i^{(j)}) + \frac{(x_i - x_i^{(j)})}{x_i^{(j+1)} - x_i^{(j)}} \cdot (f_i(x_i^{(j+1)}) - f_i(x_i^{(j)})).$$

For example, if two neighbouring $x_i^{(j)}$ values were $x_i = 4$ and $x_i = 7$, the value of the function at $x_i = 4\cdot 5$ would be approximated as

$$\hat{f}_i(4\cdot5) = f_i(4) + \frac{4\cdot5 - 4}{7 - 4}(f_i(7) - f_i(4)).$$

Similarly $g_{ki}(x_i)$ would be approximated over the same points in $x_i$ as $\hat{g}_{ki}(x_i)$, where

$$\hat{g}_{ki}(x_i) = g_{ki}(x_i^{(j)}) + \frac{x_i - x_i^{(j)}}{x_i^{(j+i)} - x_i^{(j)}}(g_{ki}(x_i^{(j+1)}) - g_{ki}(x_i^{(j)})).$$

It should be noted that both these functions are linear expressions of $x_i$ being expressed in terms of the known values of the functions and the points $x_i^{(j)}$.

These piece-wise linear functions for the objective and constraint functions can now be organized for a linear programming treatment. First we note that any value of $x_i$ lying between the points $x_i^{(j)}$ and $x_i^{(j+1)}$ can be expressed as a linear combination of $x_i^{(j)}$ and $x_i^{(j+1)}$. Using coefficients $l_{ij}$ for $x_i^{(j)}$, $x_i$ may be expressed as

$$x_i = l_{ij}x_i^{(j)} + l_{i,\,j+1}x_i^{(j+1)}$$

where $l_{ij} + l_{i,\,j+1} = 1$ and $l_{ij}$, $l_{i,\,j+1} \geqq 0$. As $l_{ij}$ and $l_{i,\,j+1}$ vary within these restrictions $x_i$ moves along the $x_i$-axis between $x_i^{(j)}$ and $x_i^{(j+1)}$. For example, the point $x_i = 4\cdot5$ where $x^{(j)} = 4$, $x^{(j+1)} = 7$ is

$$x_i = \tfrac{5}{6}.4 + \tfrac{1}{6}.7 = 4\cdot5.$$

Hence we may define $\hat{f}_i(x_i)$ as

$$\hat{f}_i(x_i) = l_{ij}f_i(x_i^{(j)}) + l_{i,\,j+1}f_i(x_i^{(j+1)})$$

which is a linear function of the $l_{ij}$ quantities.

More generally we may write for any value of $x_i$

$$x_i = \sum_{j=1}^{r} l_{ij}x_i^{(j)}$$

provided we make the restriction that there may be at most two positive $l_{ij}$ and that if there are two they are adjacent. Thus we can define $\hat{f}(x)$ over the whole range of $x$ as a function of the new variables $l_{ij}$

$$\hat{f}_i(x_i) = \sum_{j=1}^{r(i)} l_{ij}f_i(x_i^{(j)})$$

where

$$x_i = \sum_{j=0}^{r(i)} l_{ij} x_i^{(j)}$$

and

$$\sum_{j=1}^{r(0)} l_{ij} = 1, \quad l_{ij} \geq 0,$$

and we must apply the further restriction that no more than two $l_{ij}$ values are positive and that these must be adjacent for any given $i$ value. Taking the same intervals for the constraints which include functions of the $x_i$ variable we can similarly express $g_{ki}(x_i)$ in terms of the same $l_{ij}$ quantities. We thus obtain a linear programming problem in the $l_{ij}$ variables with some special restrictions on which $l_{ij}$ may be in the basis of a solution. Clearly when the method is being implemented by computer program the simplex scheme of linear programming must be rectified to incorporate these special restrictions.

The whole objective function for the sum of the separable functions becomes

$$\hat{F}(X) = \sum_{i=1}^{N} \sum_{j=1}^{r(i)} l_{ij} f_i(x_i^{(j)})$$

and the constraint functions are

$$\hat{G}_k(X) = \sum_{i=1}^{N} \sum_{j=1}^{r(i)} l_{ij} g_{ki}(x_i^{(j)}) \leq 0$$

with the additional constraints on the $l_{ij}$ quantities.

When the solution in the $l_{ij}$ variables has been obtained, these values can be used to determine the corresponding $x_i$ values by the appropriate relations. Thus the method of separable programming approximates the non-linear problem by straight line segments and enables the simplex method of linear programming to be employed in the solution.

### Example 6.7

The kind of error which may occur in a function is illustrated by approximating the function of a single variable $F(x) = x^2 - 6x + 10$ over the interval $0 \leq x \leq 5$ by the four straight line segments shown in Fig. 6.10.

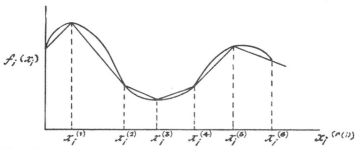

Fig. 6.10

The approximating function $\hat{F}(x)$ becomes

$$\hat{F}(x) = 10l_0 + 2l_1 + l_2 + 2l_3 + 5l_4$$

and

$$x = 0l_0 + 2l_1 + 3l_2 + 4l_3 + 5l_4,$$

where the $l$'s are subscripted only once as there is only one variable. The error at the value $x = 1$ is obtained by setting $l_0 = l_1 = \frac{1}{2}$ and

$$l_2 = l_3 = l_4 = 0.$$

Then $\hat{F}(1) = 6$ whereas $F(1) = 5$, giving an error of 1 unit.

## *6.11 Approximation programming

Approximation programming is another method for using linear programming to solve a non-linear problem. We approximate the non-linear parts of functions by using the Taylor series to obtain linear expressions which are valid in the neighbourhood of the current trial solution. We then solve the linear problem and re-approximate at the new solution. The functions can have a general form and need not, for example, be separable. The method is specially suitable when there are only a few non-linear terms in the functions.

The objective and constraint functions are separated out into their linear and non-linear terms. We will assume that there are $n$ variables $x_1, x_2, ..., x_n$ which occur linearly in both objective function and constraints and that there are $m$ variables $y_1, y_2, ..., y_m$ which occur non-linearly. Thus we wish to minimize

$$F(X, Y) = c_0 + \sum_{j=1}^{n} c_j x_j + f(y_1, y_2, ..., y_m)$$

subject to the constraints

$$G_k(X, Y) = \sum_{j=1}^{n} a_{kj} x_j + g_k(y_1, y_2, ..., y_m) = b_k \text{ for } k = 1, ..., M$$

and

$$x_j \geqq 0.$$

It is assumed that, where necessary, the constraints have been organized so that all the variables are non-negative, and that the inequalities are converted into equations by the same procedures as were used in the last chapter for converting linear problems into the standard form for linear programming.

Starting from some initial feasible solution

$$(X^{(0)}, Y^{(0)}) = (x_1^{(0)}, ... \; x_n^{(0)}, y_1^{(0)}, ..., y_m^{(0)})$$

we approximate the non-linear functions in the $y$ variables by the first two terms of the Taylor series as

$$\hat{F}(X, Y) = c_0 + \sum_{j=1}^{n} c_j x_j + f(y_1^{(0)}, y_2^{(0)}, \ldots, y_m^{(0)}) + \sum_{j=1}^{m} (y_j - y_j^{(0)}) \left[ \frac{\partial f}{\partial y_j} \right]_{Y = Y^{(0)}}$$

$$\hat{G}_k(X, Y) = \sum_{j=1}^{n} a_{kj} x_j + g_k(y_1^{(0)}, y_2^{(0)}, \ldots, y_m^{(0)}) + \sum_{j=1}^{m} (y_j - y_j^{(0)}) \left[ \frac{\partial g_k}{\partial y_j} \right]_{Y = Y^{(0)}}$$

$$= b_k, \text{ for } k = 1, \ldots, M.$$

This approximation enables the objective function to be rewritten as

$$\hat{F}(X, Y) = c_0 + \sum_{j=1}^{n} c_j x_j + c_0' + \sum_{i=1}^{m} c_i' y_i$$

where $c_i'$ are constants. For example, $c_1 = \left[ \dfrac{\partial f}{\partial y_1} \right]_{Y = Y^{(0)}}$. The constraints can

also be rewritten as

$$\sum_{j=1}^{n} a_{kj} x_j + \sum_{j=1}^{m} a_{kj}' y_j = b_k'$$

in which $a_{kj}'$, $b_k'$ are constants.

So far we have imposed no constraints on the $y_j$ variables. However, it is important that the $y_j$ values should not stray too far from $y_j^{(0)}$ before we make a new approximation. We therefore impose the limits of movement on $y_j$ as $l_j$:

$$-l_j \leqq y_i - y_j^{(0)} \leqq l_j.$$

Furthermore, by rewriting the $y_j$ variables as

$$z_j = y_j - y_j^{(0)} + l_j$$

and putting $z_j$ into the objective function and constraint equations we obtain a linear programming problem with

$$0 \leqq z_j \leqq 2l_j$$

which has the standard form with non-negative variables.

The linear programming problem is now solved and yields an answer $(X^{(1)}, Y^{(1)})$. We now form new approximations about the point $Y^{(1)}$ and repeat the whole procedure. It is important to keep the $l_j$ values small enough to ensure convergence to the actual optimum, but if they are too small this will mean that a lot of linear programming problems have to be solved.

*Example 6.8*

Maximize $F(x_1, x_2) = 2x_1 + x_2$

subject to $x_1^2 + x_2^2 \leqq 25$

$$x_1^2 - x_2^2 \leqq 7$$

$$x_1, x_2 \geqq 0.$$

The problem is illustrated in Fig. 6.11 which shows that the solution is at the point (4.3).

*First iteration*

Taking the point (1, 1) as an initial point, the first linear programming problem is:

Maximize $2x_1 + x_2$

subject to $2(x_1 - 1) + 2(x_2 - 1) \leqq 23$   or $2x_1 + 2x_2 \leqq 27$

$2(x_1 - 1) - 2(x_2 - 1) \leqq 7$   or $2x_1 - 2x_2 \leqq 7$

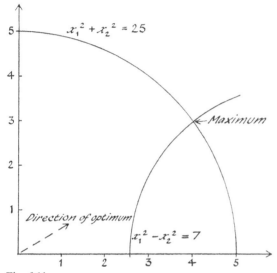

Fig. 6.11

The movement of the $x_1$, $x_2$ variables will be restricted so that

$-1 \leqq x_1 - 1 \leqq 1$

$-1 \leqq x_2 - 1 \leqq 1$.

The solution can be found graphically from Fig. 6.12 to be (2, 2).

*Second iteration*

Now approximating around the point (2, 2), the linearized problem is:

Minimize $2x_1 + x_2$

subject to $4x_1 + 4x_2 \leqq 33$

and    $4x_1 - 4x_2 \leqq 7$.

Again $-1 \leqq x_1 - 2 \leqq 1$

$-1 \leqq x_2 - 2 \leqq 1$.

This gives the solution (3, 3) as illustrated in Fig. 6.13.

It is clear how we are working towards the optimum. The remaining iterations will be done in an exercise.

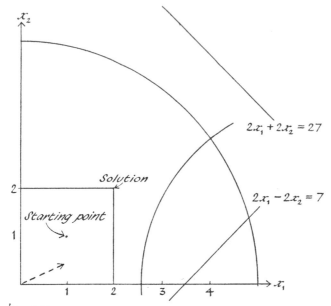

Fig. 6.12

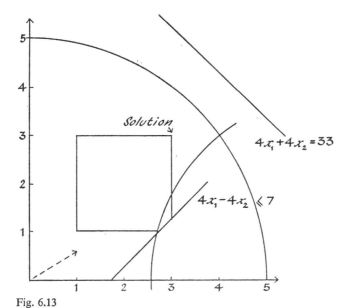

Fig. 6.13

It is worth noting that the approximation programming method can lead
to difficulties. The solution to the approximating problem may not be feasible.

For example, in Fig. 6.13 the approximating line $4x_1 + 4x_2 = 33$ acting for the constraint $x_1^2 + x_2^2 \leq 25$ is well outside the actual constraint. It is only because of the restrictions on the movement of $x_1$ and $x_2$ at an iteration 'boxing' the solution into the feasible region which prevents a non-feasible point being reached. A loss of feasibility may not matter if the next iteration forces the solution back into the feasible region. But it may be worth checking the successive linear programming solutions for feasibility in terms of the original constraints. If a non-feasible solution is obtained we can re-start from the previous starting point with tighter restrictions on the allowable variation in the non-linear variables.

# Appendix

## *6.12 The direction of steepest descent

We will now derive the direction of steepest descent which was stated in Section 6.5. Let $D = (d_1, d_2, \ldots, d_N)$ be the components of a unit vector pointing in any direction. Then if we move in the direction $D$, from the point $X = (x_1, x_2, \ldots, x_N)$, the rate of change in the function $F(x_1, x_2, \ldots, x_N)$ is given by the sum of the products of $d_i$ and $\dfrac{\partial F}{\partial x_i}$, (the rate of change of $F$ in the direction $x_i$), where the partial derivatives are evaluated at the point $X$. Hence the rate of change of the function in the direction $D$ is

$$h(d_1, d_2, \ldots, d_N) = \sum_{i=1}^{N} \frac{\partial F}{\partial x_i} \cdot d_i.$$

Now in moving from the point $X$ we wish to determine the vector

$$(d_1, d_2, \ldots, d_N)$$

such that $h$ is a minimum subject to the constraint that $D$ is a unit vector:

$$\Sigma d_i^2 = 1.$$

This minimization problem can be solved by calculus using a Lagrange multiplier. Introducing a Lagrange multiplier $y$ we take the partial derivatives of

$$h(d_1, d_2, \ldots, d_N) - y \left( \sum_{i=1}^{N} d_i^2 - 1 \right),$$

and equate to zero giving

$$\frac{\partial h}{\partial d_i} - 2yd_i = \frac{\partial F}{\partial x_i} - 2yd_i = 0 \text{ for } i = 1, \ldots, N.$$

Therefore

$$d_i = \frac{1}{2y} \cdot \frac{\partial F}{\partial x_i}.$$

By using the constraint that $\Sigma d_i^2 = 1$, we obtain

$$4y^2 = \left\{ \Sigma \left( \frac{\partial F}{\partial x_i} \right)^2 \right\}$$

giving the components $d_i$ as

$$d_i = \pm \frac{\partial F}{\partial x_i} \Big/ \left\{ \sum_{i=1}^{N} \left( \frac{\partial F}{\partial x_i} \right)^2 \right\}^{\frac{1}{2}}.$$

These two possible solutions correspond to the maximum rate of increase and the minimum rate. The direction of steepest descent is given by

$$d_i = - \frac{\partial F}{\partial x_i} \Big/ \left\{ \sum_{i=1}^{N} \left( \frac{\partial F}{\partial x_i} \right)^2 \right\}^{\frac{1}{2}}.$$

The values with the opposite signs give the direction of steepest ascent.

## *6.13  Construction of the projection matrix

In Section 6.6 on the projected gradient method it was stated that the matrix $P = I - M(M'M)^{-1}M'$ projects the direction vector $D$ along the tangent to any of the planes represented by any binding constraint of $AX \leq B$ contained in the matrix $M$. To prove this result we first note that two vectors $X$ and $Y$ are orthogonal (i.e. at right angles) to one another if their product $X.Y = 0$ and that the outward normal to the plane represented by $\sum_{j=1}^{M} a_{ij}x_j = b_i$ has direction $(a_{i1}, a_{i2}, ..., a_{iN})$. Denoting the projected direction vector by $R$, this vector must be perpendicular to the normals (i.e. parallel to the tangents). We therefore require for the matrix $M$ whose columns are the relevant rows of $A$

$$M'.R = 0.$$

The steepest descent vector $D$ may be resolved as a sum of two vectors

$$D = R + S$$

where $S$ is another vector expressing the difference between $D$ and $R$. Neglecting consideration of the rank of $M$, $S$ may be expressed in terms of the normals to the constraints. If there are $r$ columns in $M$, this is expressed as

$$S = MV$$

where $V$ is a vector of coefficients $(v_1, v_2, ..., v_r)$ giving the weights in the different directions.

Then $D = R + MV$

and premultiplying by $M'$ we get

$$M'.D = M'.R + M'.M.V$$
$$= M'MV.$$

Hence $V = (M'M)^{-1}M'.D$.

Therefore $R = D - MV$

$$= (I - M(M'M)^{-1}M)D.$$

Thus the vector $D$ is projected by the matrix

$$P = I - M(M'M)^{-1}M'$$

and the coefficients of $V$ corresponding to the various outward normals are given by

$$V = (M'M)^{-1}M'D,$$

as above.

If any of these coefficients is negative it means that the constraint is no longer binding and the constraint can be dropped.

## *6.14 Convergence of the response surface technique

We will now prove the theorem which states that the response surface technique described in Section 6.7 converges to the optimum under controls on the sequence of $r$ values chosen over the iterations. We will only consider the case for inequality constraints. The theorem states that

$$\lim_{r^{(k)} \to 0} \min_X P(X, r^{(k)}) = \min_X F(X)$$

where

$$P(X, r) = F(X) + r \sum_{k=1}^{M} 1/G_k(X)$$

and $\min_X F(X)$ denotes the minimum of $F(X)$ with respect to the variables $X$.

For the purpose of the proof we will assume that the constraints are expressed in the form $G_k(X) \geq 0$ so as to consider positive quantities. We assume that $P(X, r)$ has a unique minimum in $X$. First, we note that we are only concerned with establishing the proof when the minimum lies on the edge of the feasible region, as otherwise $P(X, r)$ becomes equivalent to $F(X)$ as $r$ becomes small. Also it is clear that for any two values of $r$, say $r^{(a)}$ and $r^{(b)}$ with $r^{(a)} < r^{(b)}$, we have

$$\min_X P(X, r^{(a)}) < \min_X P(X, r^{(b)})$$

which is proved by evaluating $P(X, r^{(b)})$ at that $X$ which minimizes $P(X, r^{(a)})$.

Let $V$ denote the minimum of $F(X)$ in the feasible region. To prove the theorem it is sufficient to show that for some arbitrarily small quantity $\varepsilon > 0$ there is an iteration number $j$ such that for $k > j$

$$\left| \min_X P(X, r^{(k)}) - V \right| < \varepsilon.$$

Since $\min\limits_{X} F(X) = V$, we have, for $X$ sufficiently close to the minimum,
$F(X) < V + \frac{1}{2}\varepsilon$.

Let the point $\overline{X}$ satisfy this condition and also lie strictly within the feasible region. Choose $r^{(j)}$ such that the following inequality is satisfied:

$$\frac{r^{(j)}}{\min\limits_{i} G_i(\overline{X})} < \frac{\varepsilon}{2M}$$

then for $k > j$

$$\min\limits_{X} P(X, r^{(k)}) < \min\limits_{X} P(X, r^{(j)})$$

$$\leqq F(\overline{X}) + r^{(j)} \sum_{i=1}^{M} \frac{1}{G_i(\overline{X})}$$

$$\leqq V + \frac{\varepsilon}{2} + r^{(j)} \sum_{i=1}^{M} \frac{1}{\min G_i(\overline{X})}$$

$$\leqq V + \frac{1}{2}\varepsilon + \frac{1}{2}\varepsilon$$

$$\leqq V + \varepsilon.$$

Also $\min\limits_{X} P(X, r^{(k)}) > V - \varepsilon$, since the penalty factor is always positive. Therefore we will come arbitrarily close to $V$ as the iterations continue. This completes the proof.

REFERENCES

Beale, E. M. L. 1967. Non-linear programming. Chapter 4.4 of *Digital Computer Users Handbook*. Eds M. Klerer and G. A. Korn. McGraw-Hill, New York.

Fiacco, A. V. and McCormick, G. P. 1964. Computational algorithm for the sequential unconstrained minimization technique for non-linear programming. *Management Sci.*, **10**, 601.

Fiacco, A. V. and McCormick, G. P. 1967. The slacked unconstrained maximization technique for convex programming. *J. Soc. Indust. Appl. Math.*, **15**, 505.

Griffith, R. E. and Stewart, R. A. 1961. A non-linear programming technique for the optimization of continuous processing systems. *Management Sci.*, **7**, 379.

Hadley, G. 1964. *Non-linear and Dynamic Programming*. Addison-Wesley, New York.

Hooke, R. and Jeeves, T. A. 1961. Direct search solution of numerical and statistical problems. *J. Assoc. Comp. Mach.*, **8**, 212.

Powell, M. J. D. 1968. A survey of numerical methods for unconstrained optimization. *Soc. Ind. Appl. Math. National Meeting*, 1968.

Rosen, J. B. 1960. The gradient projection method for non-linear programming. Part I: Linear constraints. *J. Soc. Indust. Appl. Math.*, **8**, 191.

Saaty, T. L. and Bram, J. 1964. *Non-linear Mathematics*, Chapter 3. McGraw-Hill, New York.

## Exercises on Chapter 6

**1** Determine unit vectors corresponding to the directions of steepest descent for the two variable function

$$F(x_1, x_2) = x_1^3 + 3x_1^2 - 2x_1 x_2^2 + 6$$

at the points $(1, 2)$, $(1, 1)$ and $(2, 1)$.

**2** Minimize the function of one variable

$x^2 - 7x + 9$

to an accuracy in the variable of 0·5, by the method of steepest descent, taking the origin as the initial point.

**3** It was shown by calculus methods that the function

$F(x_1, x_2) = x_1^2 + x_1 x_2 + x_2^2 + x_1^3 + x_1^2 x_2 + x_2^3$

had a minimum at the origin. Show that the method of steepest descent starting from the point $(1, 1)$ will not determine the minimum in one step regardless of its length.

**4** Demonstrate two iterations (performing one projection) of the projected gradient method on the problem:

Minimize $(x_1 - 5)^2 + (x_2 - 4)^2$

subject to $2x_1 + x_2 \leq 10$

$\qquad x_1 \qquad \leq 4$

starting with the initial interior point $(0, 4)$, and assuming that we always go up to the boundary of the next constraint.

**5** Continue with a third iteration of the problem of Exercise 4 by determining the new direction in which to move. Draw a diagram to illustrate the iteration and state what should be done to terminate the iterations.

**6** Express the created response surface $P(X, r)$ for the minimization problem:

Minimize $x_1^2 + 2x_1 x_2 + x_2^3$

subject to $14 - 3x_2 \geq 5x_1$

$\qquad x_1 + x_2 \leq 7$

$\qquad x_1^2 + x_2^2 = 3.$

**7** Sketch the form of the response surface for the problem:

Minimize $x$

subject to $x = 2$

$\qquad x \geq 0$

for $r = 1$ and $r = \frac{1}{16}$. Also show by the calculus that as $r$ approaches zero the response surface approaches its minimum for $x = 2$.

**8** Use the direct search method to minimize the function

$F(x_1, x_2) = 2x_1^2 - 7x_1 - 4x_1 x_2 - 3x_2 + 3x_2^2 + 50$

starting from the origin and using a unit step length in both variables.

Determine the exact location of the minimum by calculus, and assess what the step length of the direct search method should be reduced to in order to reach the exact optimum.

*9 For the method of separable programming determine an approximating function using at most four segments for the separable function in two variables

$$F(x_1, x_2) = x_1^3 - 6x_1^2 + 9x_1 + 2x_2 + 3$$

where $0 \leq x_1 \leq 5$

and $1 \leq x_2 \leq 3$.

*10 Continue with a further iteration of the approximation programming method of Example 6.8 with the same restrictions on the variable movement.

*11 Suggest a simple means of varying the value of $l$ in the steepest descent method over the iterations by recalculating an initial value based on the step actually taken at the previous iteration.

*12 By considering the Taylor series expansion of the function $F(X)$ at point $X^{(0)} = (x_1^{(0)}, x_2^{(0)}, \ldots, x_N^{(0)})$ in the direction of steepest descent, suggest an automatic way of determining the step length in the steepest descent method without having to conduct a search in the descent direction. Illustrate it on the problem of Example 6.1.

*13 Illustrate graphically what would happen if the projected gradient method were used on the linear programming problem discussed in Chapter 5, Example 5.4 starting from the point (0, 2). Consider briefly whether it is preferable to use the projected gradient method on linear programming problems or the simplex method.

*14 Draw a system of contours in two dimensions of a non-linear optimization problem which would lead the descent track to a boundary constraint and then back into the interior of the feasible region using the projected gradient method.

*15 Suggest a means of automatically determining the points on the $x$-axes for dividing the line segments in the method of separable programming.

# 7 Dynamic programming

## 7.1 The aim of dynamic programming

Dynamic programming is a technique for solving a special class of optimization problems called multi-stage decision processes. Unlike the previous optimization methods which have been studied, we cannot present a specific mathematical form for the class of optimization problems which dynamic programming can solve; dynamic programming is rather a computational technique for reducing the dimensionality of a problem by embodying the solution in a special set of relationships called recurrence relations. Indeed, the whole method might well have been named recursive optimization. Generally the achievement of dynamic programming is to reduce a difficult problem in $N$ variables into a series of optimization problems in one variable which are comparatively easy to solve. The reduction in computation may be quite dramatic, enabling a problem to be solved which could not otherwise be tackled. The possibility of applying the dynamic programming method depends on a successful formulation of the problem in terms of a multi-stage decision process. We will begin by describing the structure of such a process in terms of stages and states, and then we will show how this procedure can be applied to a variety of suitable problems. It is not an easy technique to present. In its completely general form, dynamic programming can seem quite meaningless, whereas an oversimplified treatment obscures its potential. In the following sections we try to strike a balance between these extremes.

## 7.2 Multi-stage decision processes

The solution to many industrial problems can be viewed in terms of taking a series of decisions which are loosely interdependent. For example the levels of production in successive weeks consist of a sequence of decisions, but the cost of production in one week may depend only on how different the current level is from the level of the previous week. Or in a chemical plant consisting of a heater, a reactor and a distillation tower connected in series, the optimal temperature, reaction rate and number of trays in the distillation tower are all interdependent, but their ideal values may be determined sequentially.

We will characterize these multi-stage processes abstractly in terms of stages and states. At each stage a decision is to be made. A stage for example may correspond to each week in a production process, or each piece of equipment in a plant. The states at the stage correspond to the alternative decisions which could be made at the stage and these will often be a range of possible values for a control variable. For the moment we will restrict ourselves to the assumption that the decision at each stage simply involves a single control variable. At each stage, therefore, the process may enter one of a number of possible states, and the process moves on from stage to stage in a sequential fashion. Fig. 7.1 illustrates a series of three stages there being 3 possible states in the first stage, 4 in the second and 2 at the third.

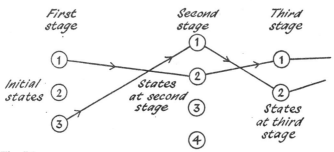

Fig. 7.1

Two possible sequences of decisions through the three stages are marked in the diagram, one leading from state 3 to state 1 to state 2 and the other leading from state 1 to state 2 to state 1.

The choice of the states at the various stages determines the effectiveness of the decisions. If the objective is expressed in terms of a minimization we associate a cost with moving from one state of a given stage to another state of the next stage. (Equally if the objective is a maximization problem we can think of a return rather than a cost.) Thus we wish to determine the optimal sequence of states to enter over the various stages. Suppose there are $N$ stages in the process and that at stage $k$ there are $M(k)$ possible states numbered

1, ..., $M(k)$. A solution to the problem may be denoted by the sequence

$$(i_1, i_2, ..., i_N)$$

where $i_k$ is the state chosen at the $k$th stage, and

$$1 \leqq i_k \leqq M(k).$$

(Often we will refer to a sequence of decisions $(i_1, i_2, ..., i_N)$ as a policy.) If $F(i_1, i_2, ..., i_N)$ denotes the total cost associated with the sequence

$$(i_1, i_2, ..., i_N),$$

the formal optimization problem is to minimize $F$ subject to any constraints which are placed on the choice of the states at each stage, and these constraints may depend on the states entered at stages 1, 2, ..., $k-1$. This is an optimization problem in $N$ dimensions. We will now describe how in suitable circumstances it can be turned into a series of problems in one variable using dynamic programming.

## 7.3  The dynamic programming method: forwards calculation

There are two ways of applying the dynamic programming method depending on whether the calculation goes forward through the successive stages 1, ..., $N$ or works backwards starting from the last stage and proceeding in the reverse direction to stage 1. Although dynamic programming is most commonly presented in terms of the backward calculation we will postpone a discussion of it until the forwards direction has been described as this is the simpler way to obtain an understanding of the ideas.

We wish to determine the sequence of states $i_1, i_2, ..., i_N$ for the $N$-stage decision problem so as to minimize $F(i_1, i_2, ..., i_N)$. The dynamic programming method is based on the intuitively obvious assertion that an optimal sequence of decisions has the property that, at any stage, whatever the next state and next decision are, the previous decisions must constitute an optimal policy leading up to this state and stage. We can use this assertion to relate the best policy leading up to state $i_k$ at stage $k$ to the best policy up to the alternatives at stage $(k-1)$.

We define $f_k(i_k)$ as the minimum possible cost of the $k$-stage process leading to state $i_k$ at stage $k$. This apparently innocuous looking definition is often the key to the successful solution of a problem by dynamic programming. Then if the policy $(i_1, i_2, ..., i_k)$ is to be optimal we require that the policy up to state $i_{k-1}$ be optimal. Let us denote by

$$G((i_k, i_{k-1}, f_{k-1}(i_{k-1}))$$

the cost of moving through state $i_{k-1}$ of stage $k-1$ to state $i$ of stage $k$. Then the condition for an optimal policy stated above asserts that we must choose the $i_{k-1}$ value so as to minimize the function $G$. This leads to a recurrence relation based on the following relation:

Cost to $i_k$ = minimum for choice of $i_{k-1}$ (cost to $i_k$ via $i_{k-1}$) or, formally,

$$f_k(i_k) = \min_{1 \leq i_{k-1} \leq M(k-1)} G[i_k, i_{k-1}, f_{k-1}(i_{k-1})]$$

where $M(k-1)$ is the number of states available at stage $(k-1)$ assumed to be numbered 1, 2, ..., $M(k-1)$.

Usually the function $G$ will be a simple combination of the direct cost of the transfer say $c(i_{k-1}, i_k)$ between the two states and the total cost up to this point $f_{k-1}(i_{k-1})$. Typically we then have simple recurrence relations of the form:

$$f_k(i_k) = \min_{1 \leq i_{k-1} \leq M(k-1)} [c(i_{k-1}, i_k) + f_{k-1}(i_{k-1})]$$

or

$$_k(i_k) = \min_{1 \leq i_{k-1} \leq M(k-1)} [c(i_{k-1}, i_k) \cdot f_{k-1}(i_{k-1})].$$

We solve the recurrence equation for all values $i_k = 1, ..., M(k)$ and for $k = 1, ..., N$, and the optimal value of the whole policy is:

$$\min_{1 \leq i_N \leq M(N)} f_N(i_N)$$

as this is the least cost of moving from the first stage to the last stage and choosing the best final state. This gives us the objective function value, but it does not determine the optimal policy or sequence $(i_1, i_2, ..., i_N)$ which leads to this value. However, while solving the recurrence relations we can record the value of $i_{k-1}$ for which the minimum is achieved for $f_k(i_k)$. Let us denote this minimum index by $q_{k-1}(i_k)$.

Thus

$$f_k(i_k) = \min_{1 \leq i_{k-1} \leq M(k-1)} G[i_k, i_{k-1}, f_{k-1}(i_{k-1})] = G[i_k, q_{k-1}(i_k), f_{k-1}(q_{k-1}(i_k))].$$

Then the optimum policy is determined by working backwards through the sequence as follows:

Given $\bar{i}_N$ is the state selected at the final stage,

$\bar{i}_{N-1} = q_{N-1}(i_N)$ is the state at stage $N-1$,

$\bar{i}_{N-2} = q_{N-2}(q_{N-1}(\bar{i}_N))$ is the state at stage $N-2$,

$\bar{i}_{N-3} = q_{N-3}(q_{N-2}(q_{N-1}(\bar{i}_N)))$ is the state at stage $N-3$,

$$\vdots$$

and in general if $\bar{i}_k$ is the state at stage $k$,

$\bar{i}_{k-1} = q_{k-1}(\bar{i}_k)$ is the state at stage $k-1$.

The optimal policy is therefore $(\bar{i}_1, \bar{i}_2, ..., \bar{i}_N)$, and it is clear how the forwards calculation of the dynamic programming procedure in fact requires a 'backwards pass' to determine the optimal policy.

The method has been presented without reference to constraints. The dynamic programming procedure can readily handle constraints on the set of states from which we can move at the next stage. In general, let $S_{k-1}(i_k)$ denote the set of states at stage $(k-1)$ from which we can move to state $i_k$ at stage $k$. Then the recurrence relations may be written to include these constraint considerations as

$$f_k(i_k) = \min_{i_{k-1} \in S_{k-1}(i_k)} G[i_k, i_{k-1}, f_{k-1}(i_{k-1})].$$

There is no difficulty about handling constraints of this kind, and in fact they reduce the amount of computation which has to be carried out. This constraint handling facility is one of the main virtues of the dynamic programming method.

## 7.4 The backwards calculation

The original presentation of dynamic programming by Bellman was based on the principle of optimality which stated that 'an optimal policy has the property that whatever the initial state and initial decision are, the remaining decisions must constitute an optimal policy with regard to that decision'. This is the reverse of the optimality assertion made for the forwards calculation presented in the previous section. Here we are stating that if we are currently in state $i_k$ of stage $k$ and there are $(N-k)$ decisions still to make, we must determine an optimal policy for the remaining decisions for the whole policy to constitute an optimal policy. Therefore we define $f_{N-k}(i_k)$ as the minimum cost or optimal return which can be obtained over the stages

$(k+1)$, $(k+2)$, ..., $N$.

Notice that this function $f_{N-k}(i_k)$ is proposed without the decisions yet having been made. Then if $G_k(i_k, i_{k+1}, f_{N-k-1}(i_{k+1}))$ is the cost of going from state $i_k$ to $i_{k+1}$ at stage $(k+1)$, the principle of optimality yields the recurrence relations:

$$f_{N-k}(i_k) = \min_{i_{k+1} \in S_{k+1}(i_k)} G(i_k, i_{k+1}, f_{N-k-1}(i_{k+1}))$$

where $S_{k+1}(i_k)$ is the set of eligible states to move to from $i_k$. This system of equations is solved backwards for $k = N, N-1, N-2, ..., 1$ starting at stage $N$ and ending at stage 1. When the optimal objective function value $f_{N-1}(\bar{i}_1)$ is determined the policy is found in a forwards pass over the stages as the sequence $\bar{i}_1, \bar{i}_2, .... \bar{i}_N$.

Often a problem can be solved in the forwards or backwards directions equally easily and it is then usually simpler to adopt the forwards approach. However, there may be substantial conceptual and computational returns in viewing the problem from one end rather than another. The backwards approach is particularly useful when we know the terminal state at the final stage which we wish to reach. This type of situation arises for example in chemical process control where we know the final composition which is required and the dynamic programming calculation determines how to reach it. The backwards calculation is undoubtedly more subtle and once its meaning has been exactly understood it more properly represents the potential of the dynamic programming method.

The only way in which dynamic programming can really be understood is to see how optimization calculations can be formulated as recurrence relations, and to practise expressing new problems in dynamic programming terms. Three typical areas for the application of dynamic programming will now be investigated by solving particular examples. The applications are in network routing, resource allocation and reliability of multi-component devices. All of them will be solved by the forwards calculation, but the backwards calculation could equally well apply.

### 7.5 Network problems

A very wide range of problems can be reduced to the task of finding the shortest route through a network from a given starting point in the network to a specified final point. Dynamic programming is very efficient for solving this problem and we will show how it is harnessed for this task in the following example.

*Example* 7.1

The network shown in Fig. 7.2 is a set of road links connecting a number of towns. The distances between the various points are marked on the connections. The objective is to find the shortest route between points 1 and 11. First the problem is defined in terms of stages and states. As it will be necessary to proceed from point 1 through points 2, 3 or 4 followed by points 5, 6 or 7 followed by points 8, 9, or 10, we can regard point 1 as the only state of

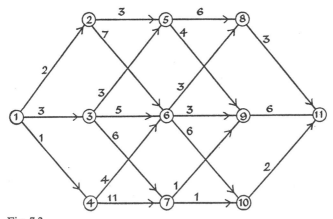

Fig. 7.2

stage 1; points 2, 3, 4 as constituting the states of stage 2; points 5, 6, 7 as stage 3; points 8, 9, 10 as stage 4; and point 11 as stage 5.

The states of each stage which can lead on to the different states of the next stage are clearly indicated by the arrows. Let $d(i, i_k)$ denote the distance from point $i$ to point $i_k$, and $f_k(i_k)$ denote the minimum distance from point 1 to point $i_k$ at stage $k$. Using the principle of optimality, we assert that the minimum distance to point $i_k$ at stage $k$ must equal the minimum distance up to some point $i$ at stage $(k-1)$ plus the distance between points $i$ and $i_{k-1}$. This gives the recurrence relations

$$f_k(i_k) = \min_{i \in S_{k-1}(i_k)} \{d(i, i_k) + f_{k-1}(i)\}.$$

where $S_{k-1}(i_k)$ are the towns which can be passed through at stage $k-1$ to reach state $i_k$. Setting $f_1(1) = 0$ the recurrence relations are evaluated in turn for $k = 1, 2, 3, 4$ and 5. Also at any stage $k$ we evaluate $f_k(i_k)$ for all states $i_k$ belonging to the stage and record in $q_{k-1}(i_k)$ the point to come from at the $(k-1)$th stage. Solving the recurrence relations for the network we obtain the results presented below:

$f_1(1) = 0$

$f_2(2) = \min_{i=1} [d(i, 2) + f_1(i)] = 2, q_1(2) = 1$

$f_2(3) = \min_{i=1} [d(i, 3) + f_1(i)] = 3, q_1(3) = 1$

$f_2(4) = \min_{i=1} [d(i, 4) + f_1(i)] = 1, q_1(4) = 1$

$f_3(5) = \min_{i=2,3} [d(i, 5) + f_2(i)] = 5, q_2(5) = 2$

$f_3(6) = \min_{i=2,3,4} [d(i, 6) + f_2(i)] = 5, q_2(6) = 4$

$f_3(7) = \min_{i=3,4} [d(i, 7) + f_2(i)] = 9, q_2(7) = 3$

$f_4(8) = \min_{i=5,6} [d(i, 8) + f_3(i)] = 8, q_3(8) = 6$

$f_4(9) = \min_{i=5,6,7} [d(i, 9) + f_3(i)] = 8, q_3(9) = 6$

$f_4(10) = \min_{i=6,7} [d(i, 10) + f_3(i)] = 10, q_3(10) = 7$

$f_5(11) = \min_{i=8,9,10} [d(i, 11) + f_4(i)] = 11, q_4(11) = 8.$

Therefore the minimum distance is 11 and the route is

$q_4(11) = 8, q_3(8) = 6, q_2(6) = 4, q_1(4) = 1$

i.e. 1, 4, 6, 8, 11.

## 7.6 Allocation of a single resource to a number of activities

Perhaps the most successful application of dynamic programming to operations research problems arises in the allocation of a single resource in limited supply to a number of independent activities. We wish to optimize the distribution of the resource by maximizing the total return. Suppose there are $N$ activities or uses to which we wish to apply a resource. The return on allocating a quantity $x_i$ of the resource to activity $i$ is $g_i(x_i)$ and the total quantity of the resource available is $S$ units. Then the problem is to determine values $x_1, x_2, \ldots, x_N$ which maximize the total return which is expressed by the objective function

$$F(x_1, x_2, \ldots, x_N) = \sum_{i=1}^{N} g_i(x_i)$$

subject to the constraints

$$x_1 + x_2 + \ldots + x_N \leq S$$

and $x_i \geq 0$ for all $i$,

where the last constraints state that it is impossible to allocate a negative amount of the resource.

To formulate this problem in dynamic programming terms we can view the allocations to the successive activities as a series of stages. The states at stage $k$ can be defined as the total quantity of the resource which has been allocated up to and including the allocation at stage $k$. Note that this is not just the amount allocated to activity $k$. We may vary this discretely determining the allocations for example in percentages of $S$ as $S \cdot i/100$ for $i = 0, \ldots, 100$ implying 101 possible states at each stage.

We define $f_k(y)$ as the maximum possible return obtainable if a total of $y$ units are allocated over the first $k$ activities. The quantity $f_k(y)$ can be related to $f_{k-1}(z)$ in the following way. Let $x_k$ denote the allocation of resource to activity $k$ at stage $k$. The function $f_{k-1}(z)$ has recorded the optimal allocation to $(k-1)$ activities if a total quantity of the resource $z$ is used. The quantity $y$ must therefore be divided between the allocation $z$ to the first $(k-1)$ activities and the allocation $x_k$ to the $k$th activity. If we make the assumption that the $g_k(x_k)$ functions are monotone increasing (i.e. the return increases as the allocation increases), then the state at stage $(k-1)$ (i.e. level of $z$) which we are interested in is simply the level $(y-x_k)$. Because of monotonicity it is always an advantage to allocate as much resource as possible.

Therefore, if $x_k$ is allocated to activity $k$ and the remaining $(y-x_k)$ units are allocated optimally over the first $(k-1)$ activities, the total return is

$$g_k(x_k) + f_{k-1}(y - x_k).$$

Since $f_k(y)$ is defined as the maximum return on allocating $y$ units to the first

$k$ activities, we must choose $x_k$ optimally to determine $f_k(y)$ as

$$f_k(y) = \max_{0 \leq x_k \leq y} [g_k(x_k) + f_{k-1}(y - x_k)]$$

where the expression is evaluated over the discrete set of values for $x_k$. If we take percentages of $S$, these will be $x_k = 0, S/100, 2S/100, ..., y$. These are the dynamic programming recurrence relations. To help to explain the recurrence relation, the expression for $f_k(y)$ may be written out in full as

$$f_k(y) = \max_{0 \leq x_k \leq y} [g_k(x_k) + \max_{\substack{x_1, x_2, ..., x_{k-1} \\ x_1 + x_2 + ..., x_{k-1} \leq y - x_k}} \{(g_1(x_1) + g_2(x_2) + ... + g_{k-1}(x_{k-1})\}].$$

To solve the problem, we need to evaluate this recurrence relation for $y = S.i/100$ for $i = 0, 1, 2, ..., 100$, and for $k = 1, ..., N$. The maximum value is $f_N(S)$. Also, each time $f_k(y)$ is determined it is necessary to record the value of $x_k$ for which the maximum was obtained. In terms of the previous notation, if $x_k$ is allocated at stage $k$, we could record the state of the previous stage which led to $x_k$ as $q_{k-1}(y) = y - x_k$. However, it is more natural here to record the actual value of $x_k$ for which the maximum occurs at stage $k$ and we will denote this by $q_k(y)$. The optimal quantities to allocate to the stages are $\bar{x}_N, \bar{x}_{N-1}, \bar{x}_{N-2}, ..., \bar{x}_1$ determined in that order as

$$\bar{x}_N = q_N(S)$$

$$\bar{x}_{N-1} = q_{N-1}(S - \bar{x}_N)$$

$$\bar{x}_{N-2} = q_{N-2}(S - \bar{x}_N - \bar{x}_{N-1})$$

$$\vdots$$

$$\bar{x}_i = q_i\left(S - \sum_{k=i+1}^{N} \bar{x}_k\right)$$

$$\vdots$$

$$\bar{x}_1 = q_i\left(S - \sum_{k=2}^{N} \bar{x}_k\right).$$

We will now give two examples of this sort of problem.

*Example 7.2*

A limited amount of 5 tons of fertilizer is to be applied to 3 crops. The return on the different levels of fertilizer for the various crops is shown in Table 7.1. It is assumed that an integral number of tons will be applied to each crop. The data in the table corresponds to the functions $g_k(y)$, the return from allocating $y$ tons to the $k$th crop.

The allocation to the separate crops corresponds to the stages and in terms of the previous notation the table gives the $g_1(x_1)$, $g_2(x_2)$, and $g_3(x_3)$ values. Define $f_k(y)$ as the maximum return which can be obtained by allocating $y$ tons to the first $k$ crops and $q_k(y)$ as the amount allocated to the $k$th crop.

Then for the first stage

$$f_1(0) = 2 \quad q_1(0) = 0$$
$$f_1(1) = 3 \quad q_1(1) = 1$$
$$f_1(2) = 3 \quad q_1(2) = 2$$
$$f_1(3) = 4 \quad q_1(3) = 3$$
$$f_1(4) = 4 \quad q_1(4) = 4$$
$$f_1(5) = 4 \quad q_1(5) = 5.$$

**Table 7.1**

| Fertilizer level in tons | Crops | | |
|:---:|:---:|:---:|:---:|
| | 1 | 2 | 3 |
| 0 | 2 | 2 | 1 |
| 1 | 3 | 2 | 2 |
| 2 | 3 | 4 | 3 |
| 3 | 4 | 4 | 6 |
| 4 | 4 | 6 | 6 |
| 5 | 4 | 8 | 6 |

For the second crop,

$$f_2(0) = \max_{x_2 = 0} (g_2(0) + f_1(0)) = 4 \qquad\qquad q_2(0) = 0$$

$$f_2(1) = \max_{x_2 = 0, 1} (g_2(x_2) + f_1(1 - x_2)) = \max(5, 4) = 5 \qquad\qquad q_2(1) = 0$$

$$f_2(2) = \max_{x_2 = 0, 1, 2} (g_2(x_2) + f_1(2 - x_2)) = \max(5, 5, 6) = 6 \qquad q_2(2) = 2$$

$$f_2(3) = \max_{x_2 = 0, 1, 2, 3} (g_2(x_2) + f_1(3 - x_2)) = \max(6, 5, 7, 6) = 7 \qquad q_2(3) = 2$$

$$f_2(4) = \max_{x_2 = 0, 1, 2\ 3, 4} (g_2(x_2) + f_1(4 - x_2)) = \max(6, 6, 7, 7, 8) = 8 \qquad q_2(4) = 4$$

$$f_2(5) = \max_{x_2 = 0, 1, 2, 3, 4, 5} (g_2(x_2) + f_1(5 - x_2)) = \max(6, 6, 8, 7, 9, 10) = 10 \quad q_2(5) = 5.$$

Similarly, the recurrence relation

$$f_3(y) = \max_{0 \le x_3 \le y} [g_3(x_3) + f_2(y - x_3)]$$

is used to determine the values for $f_3(y)$ and $q_3(y)$ for $y = 0, 1, 2, 3, 4, 5$. The results are tabulated in Table 7.2. It shows that the maximum return of 12 is obtained by applying 3 tons to crop 3, $q_2(5-3) = 2$ tons to crop 2 and $q_1(5-3-2) = 0$ tons to crop 1. It may be noticed that there are a number of alternative allocations which will all achieve this solution.

The table also provides the optimum solutions if less than 5 tons of fertilizer are available without any additional calculation. For example if only 4 tons were available the table shows that

$$q_3(4) = 3, \ q_2(4-3) = 0 \text{ and } q_1(4-3-0) = 1,$$

implying that 3 tons should be allocated to crop 3 and 1 ton to crop 1.

Thus a whole family of solutions has been obtained. Sometimes this can be an important advantage resulting from the dynamic programming method.

**Table 7.2**

| $y$ | $f_1(y)$ | $q_1(y)$ | $f_2(y)$ | $q_2(y)$ | $f_3(y)$ | $q_3(y)$ |
|---|---|---|---|---|---|---|
| 0 | 2 | 0 | 4 | 0 | 5 | 0 |
| 1 | 3 | 1 | 5 | 0 | 6 | 0 |
| 2 | 3 | 2 | 6 | 2 | 7 | 0 |
| 3 | 4 | 3 | 7 | 2 | 10 | 3 |
| 4 | 4 | 4 | 8 | 4 | 11 | 3 |
| 5 | 4 | 5 | 10 | 5 | 12 | 3 |

*Example 7.3*

A problem in cargo loading is to load a given capacity with the most valuable cargo. Let the capacity of the container be $z$ units, whose cargo consists of different quantities of $N$ different items, which are distinguished by value and weight. Let $v_i$ and $w_i$ denote the value and weight of the $i$th type of item. Then if we load $x_i$ items of type $i$, the value of the cargo which we wish to maximize is

$$f(x_1, x_2, \ldots, x_N) = \sum_{i=1}^{N} v_i x_i$$

subject to the constraints

$$\sum_{i=1}^{N} w_i x_i \leqq z$$

$$x_i = 0, 1, 2, \ldots.$$

(Notice that this problem cannot be tackled by linear programming because of the integer values for the variables $x_i$.)

If we consider the different item types to be loaded sequentially we can regard the $N$ allocations as representing the $N$ stages of the dynamic programming procedure. The states at stage $k$ are the various amounts of the total capacity which may have been occupied on the allocations up to stage $k$. We define $f_k(y)$ as the optimal value which may be obtained by allocating items of types 1 to $k$ using a total capacity of $y$. Then the recurrence relations are expressed as:

$$f_k(y) = \max_{0 \le x_k \le \left[\frac{y}{w_k}\right]} \{x_k v_k + f_{k-1}(y - x_k w_k)\}$$

where $\left[\dfrac{y}{w_k}\right]$ denotes the largest integer less than $(y/w_k)$. We evaluate this recurrence relation for $y = 1, 2, 3, \ldots$, and $k = 1, \ldots, N$. For a given $y$ value, we can record the optimum allocation as $q_k(y)$. The maximum return is $f_N(z)$ and if $\bar{x}_k$ denotes the number of items of type $k$ in the optimum allocation, $x_k$ is determined by the recurrence relations

$$\bar{x}_k = q_k\left(z - \sum_{i=k+1}^{N} w_i \bar{x}_i\right).$$

A small numerical example will help to illustrate the procedure. Suppose there is a total weight capacity of 9 and there are 4 item types whose weights and values are as shown in Table 7.3. The calculations are laid out in Table 7.4 Thus the maximum value is 25 and this is achieved by loading 1 item of type 4, 1 of type 3.

**Table 7.3**

| Item | Weight | Value |
|------|--------|-------|
| 1 | 2 | 5 |
| 2 | 3 | 8 |
| 3 | 5 | 14 |
| 4 | 4 | 11 |

**Table 7.4**

| $y$ | STAGE 1 $f_1(y)$ | $q_1(y)$ | STAGE 2 $f_2(y)$ | $q_2(y)$ | STAGE 3 $f_3(y)$ | $q_3(y)$ | STAGE 4 $f_4(y)$ | $q_4(y)$ |
|-----|------|------|------|------|------|------|------|------|
| 0 | 0 | 0 | 0 | 0 | 0 | 0 | 0 | 0 |
| 1 | 0 | 0 | 0 | 0 | 0 | 0 | 0 | 0 |
| 2 | 5 | 1 | 5 | 0 | 5 | 0 | 5 | 0 |
| 3 | 5 | 1 | 8 | 1 | 8 | 0 | 8 | 0 |
| 4 | 10 | 2 | 10 | 0 | 10 | 0 | 11 | 1 |
| 5 | 10 | 2 | 13 | 1 | 14 | 1 | 14 | 0 |
| 6 | 15 | 3 | 16 | 2 | 16 | 0 | 16 | 0 |
| 7 | 15 | 3 | 18 | 1 | 19 | 1 | 19 | 0 |
| 8 | 20 | 4 | 21 | 2 | 22 | 1 | 22 | 0 |
| 9 | 20 | 4 | 24 | 3 | 24 | 0 | 25 | 1 |

Again the dynamic programming procedure has solved a family of problems. The table contains the optimum allocations for cargoes of total capacity $y = 2, 3, 4, 5, 6, 7, 8, 9$. Table 7.5 shows the maximum cargo values for each of these capacities and the items included in each case.

**Table 7.5**

| $y$ | Cargo value | Items |
|-----|-------------|-------|
| 2 | 5 | $\bar{x}_1 = 1$ |
| 3 | 8 | $\bar{x}_2 = 1$ |
| 4 | 11 | $\bar{x}_4 = 1$ |
| 5 | 14 | $\bar{x}_3 = 1$ |
| 6 | 16 | $\bar{x}_2 = 2$ |
| 7 | 19 | $\bar{x}_1 = 1, \bar{x}_3 = 1$ |
| 8 | 22 | $\bar{x}_2 = 1, \bar{x}_3 = 1$ |
| 9 | 25 | $\bar{x}_3 = 1, \bar{x}_4 = 1$ |

## 7.7 Reliability problems

A further application of dynamic programming in the field of reliability is illustrated in the next example. The form of the functions in the recurrence relations here are rather different.

*Example 7.4*

An electronic device which is arranged as a sequence of $N$ stages is to be constructed. Each stage consists of one or more units of a given type of component as illustrated in Fig. 7.3. The reliability of each stage depends on the number of duplicate components inserted in the stage. Let $x_j$ be the number of components at the $j$th stage. (It is assumed that there is at least one component at each stage.) The probability of the $j$th stage functioning successfully is $p_j(x_j)$. Clearly $p_j(x_j)$ increases as $x_j$ increases. The stages are presumed to be independent, so that the total reliability $F(x_1, x_2, ..., x_N)$ of the system is the product of the probabiliies

$$F(x_1, x_2, ..., x_N) = p_1(x_1) \cdot p_2(x_2) ... p_N(x_N).$$

The cost of a component for the $j$th stage is $c_j$ and the problem is to determine the number of components which should be inserted at each stage so as to maximize the reliability subject to not exceeding a total cost of construction $C$. This is expressed as the constraint

$$c_1 x_1 + c_2 x_2 + ... + c_N x_N \leqq C.$$

To solve the problem by dynamic programming, we can view the stages of the device as the stages of the dynamic programming process. We define $f_k(y)$ as the maximum reliability if a total cost of $y$ is made for the devices of the first $k$ stages. $y$ is allowed to range in the interval $0 \leq y \leq C$. If $x_k$ denotes the number of components allocated to stage $k$, the optimality conditions state that the greatest reliability which can be obtained if total funds $y$ are allocated to the first $k$ stages is the product of $p_k(x_k)$ and $f_{k-1}(y - c_k x_k)$, where $x_k$ is chosen to maximize the product. This provides the recurrence relations

$$f_k(y) = \max_{0 \leq x_k \leq \left[\frac{y}{c_k}\right]} \{[p_k(x_k) \cdot f_{k-1}(y - c_k x_k)]\}$$

and $f_1(y) = p_1 \left[\dfrac{y}{c_1}\right]$, where $\left[\dfrac{y}{c_1}\right]$ is the largest integer less than $(y/c_1)$. We can record the best number of components to allocate to stage $k$ if a total cost of $y$ is assigned to the first $k$ stages as $q_k(y)$. Then the optimal reliability of the whole system is

$$\max_{0 \leq x_N \leq C} \{p_N(x_N) \cdot f_{N-1}(C - c_N x_N)\}$$

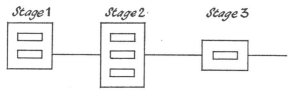

Fig. 7.3

and the number of components to assign to each stage is determined as usual by the backwards pass giving

$$q_N(C), \; q_{N-1}(C - c_N \cdot q_N(C)), \; \ldots.$$

## 7.8 Computational comments

The real achievement of dynamic programming is a great reduction in computation. Let us consider the amount of calculation involved in an allocation problem. Suppose there are $N$ stages and $M(k)$ is the number of states at stage $k$. The dynamic programming calculation determines the optimum by

solving $\sum_{k=1}^{N} M(k)$ minimization problems in one variable. The original minimization involved one problem in $N$ variables which in general is a much harder task. Furthermore, it is certain that the global optimum solution will be obtained.

It is interesting to look at the detailed computation involved in the allocation problems. Suppose each stage has $M$ states. Then each stage requires the calculation of $M$ values $f_k(y)$. The determination of each $f_k(y)$ value requires $M$ additions and $M$ comparisons. Therefore the total number of additions and comparisons are both $N.M^2$. It should be noted that the amount of computation goes up linearly with $N$ but in proportion to the square of the number of states, so that adding extra stages to the allocation problem will not increase the computation excessively whereas increasing the number of states will increase the amount of calculation substantially. If a direct evaluation were made of all possible solutions this would involve much more work. There are $N^M$ possible solutions and each of these involves $N$ additions. $N^M$ is a very much larger number than $N.M^2$. If there were 10 states and 10 stages direct evaluation requires ten thousand million additions and comparisons compared with the thousand required by dynamic programming.

However, the dynamic programming method also runs into difficulties on what appear to be quite simple problems when the number of states at the various stages becomes large. We will demonstrate this on an allocation problem involving two resources in Section 7.9. There are a large number of multi-stage problems which can be formulated in dynamic programming terms but which cannot possibly be solved because of the computational requirements.

The computational schemes which have been presented have always performed the minimization or maximization in the recurrence relation by direct evaluation. This is not always the best approach. Once the problem is expressed in the form of a single variable minimization problem any optimization method can be used to solve the problem. Sometimes it is possible to use the calculus as illustrated in the following example.

*Example 7.5*

A positive quantity $C$ is to be divided into three parts so that the product of the three parts is a maximum. Let $x_1$, $x_2$, $x_3$ denote the values of the parts, and define $f_k(y)$ as the maximum product if a quantity $y$ is divided into $k$ parts. Then, as $y$ must be composed of the $k$th part $x_k$ and the remainder $(y - x_k)$ optimally divided into $(k - 1)$ parts,

$$f_k(y) = \max_{0 \le x_k \le y} (x_k . f_{k-1}(y - x_k)),$$

setting

$$f_1(y) = y,$$
$$f_2(y) = \max_{0 \le x_2 \le y} (x_2 . (y - x_2)),$$

Now the function

$$g_2(x_2) = x_2(y - x_2)$$

can be differentiated to give a maximum solution:

$$\frac{dg_2}{dx_2} = y - 2x_2 = 0$$

giving $x_2 = y/2$ as the maximum.

E

Thus $f_2(y) = \dfrac{y}{2}\left(y - \dfrac{y}{2}\right) = \left(\dfrac{y}{2}\right)^2.$

Similarly

$$f_3(y) = \max_{0 \le x_3 \le y} \ (x_3 \cdot f_2(y - x_3))$$

$$= \max_{0 \le x_3 \le y} \ \left(x_3 \left(\dfrac{y - x_3}{2}\right)^2\right).$$

Now if $g_3(x_3) = x_3 \cdot \left(\dfrac{y - x_3}{2}\right)^2$

$$\dfrac{dg_3}{dx_3} = -x_3\left(\dfrac{y - x_3}{2}\right) + \left(\dfrac{y - x_3}{2}\right)^2 = 0$$

giving $x_3 = y/3$ as the maximum.

Therefore $f_3(y) = \left(\dfrac{y}{3}\right)^3$ and the maximum value is $\left(\dfrac{C}{3}\right)^3$, each part being $C/3$.

## *7.9  Allocation processes involving two types of resources or two constraints

A natural extension of the allocation process of Section 7.6 is to consider two different types of resources being applied to a number of independent activities. Following an analogous formulation to the single resource problem, suppose there are $N$ activities and the supplies of the two types of resources are $A$ and $B$. Let $x_i$ and $y_i$ be the quantities of the two resources allocated to activity $i$ and let $g_i(x_i, y_i)$ denote the corresponding return from the $i$th activity. Then we wish to maximize the function in $2N$ variables:

$$F(x_1, x_2, \dots, x_N; y_1, y_2, \dots, y_N) = \sum_{i=1}^{N} g_i(x_i, y_i),$$

subject to the constraints

$$\sum_{i=1}^{N} x_i \le A,$$

$$\sum_{i=1}^{N} y_i \le B,$$

$$x_i \ge 0,$$

$$y_i \ge 0.$$

We define the function $f_k(a, b)$ as the maximum return which can be obtained from allocating a total of $a$ units of the first resource and $b$ units of the second resource to the first $k$ activities. Thus

$$f_k(a, b) = \max \sum_{i=1}^{k} g_i(x_i, y_i),$$

where the maximization is taken over $x_1, x_2, ..., x_k$ and $y_1, y_2, ..., y_k$ subject to

$$\sum_{i=1}^{k} x_i \leq a \quad \text{and} \quad \sum_{i=1}^{k} y_i \leq b.$$

We thus obtain the recurrence relations:

$$f_k(a, b) = \max_{\substack{0 \leq x_k \leq a \\ 0 \leq y_k \leq b}} \{f_k(x_k, y_k) + f_{k-1}(a - x_k, b - y_k)\}.$$

So far this appears straightforward. These relations would need to be evaluated for $0 \leq a \leq A$, $0 \leq b \leq B$ and for $k = 1, ..., N$, and the problem is solved in exactly the same way as for the one resource case. But it should be noticed here that the number of states which have to be recorded has increased greatly, as they now have to be recorded over a square region rather than along a line. This will mean that the problem can only be solved for comparatively small ranges of state variables.

The same difficulties arise if there are two constraints acting on the allocation of a single resource. Suppose we wish to maximize the function

$$F(x_1, x_2, ..., x_N) = g_1(x_1) + g_2(x_2) + ... + g_N(x_N)$$

subject to the constraints

$$a_1 x_2 + a_2 x_2 + ... + a_N x_N \leq A$$
$$b_1 x_1 + b_2 x_2 + ... + b_N x_N \leq B$$
$$x_i \geq 0.$$

We have to consider the allocation of the quantities $A$ and $B$ separately. A new function $f_k(X, Y)$ is defined which represents the maximum return which can be obtained over the first $k$ functions $g_i(x_i)$ subject to the restriction that

$$\sum_{i=1}^{k} a_i x_i \leq X$$

and

$$\sum_{i=1}^{k} b_i x_i \leq Y.$$

Then the recurrence relations become

$$f_k(X, Y) = \max_{\substack{0 \leq a_k x_k \leq X \\ 0 \leq b_k x_k \leq Y}} \{g_k(x_k) + f_{k-1}(X - a_k x_k, Y - b_k x_k)\}.$$

Again this function must be evaluated over a square region, so that the computing time required increases quickly as the state space increases. It emphasizes the assertion made in Section 7.8 that the computational requirements increase severely as the number of states gets larger. Sometimes it is possible to overcome some of the difficulties by using a Lagrange multiplier to build equality constraints into the objective function. But too much computation due to a large number of states is usually the threat from further constraints of this type. It is a manifestation of what Bellman has called 'the curse of dimensionality'.

## *7.10  Problems with two neighbouring stages related: backward calculations

In the previous problems, it has been possible to employ the system of the forward calculation described in Section 7.3 in which the recurrence equations relate the functions $f_k$ of stage $k$ to the functions $f_{k-1}$ of stage $(k-1)$. In other problems it is more natural to use the backward system discussed in Section 7.4 where the function $f_k$ defines the optimal return obtainable from the remaining $(N-k)$ stages. This is sometimes the case when the return at stage $k$ depends on the state selected at stage $(k-1)$ and each pair of neighbouring stages are related. Problems of this type arise in such widely differing fields as stock control and statistical mechanics. We will now consider the backwards calculations on a problem of this type, although it is also straightforward to tackle it by a forwards approach.

Consider a problem in which we wish to determine values $x_1, x_2, ..., x_N$, to maximize the function

$$F(x_1, x_2, ..., x_N) = g_1(x_1 - x_0) + g_2(x_2 - x_1) + ... + g_N(x_N - x_{N-1}).$$

We will assume that all $x_i \geq 0$, and $x_N$ is given as a terminal value required for the process.

The maximization of the function $F$ can be split up into stages such as $x_1$ and the remainder as:

$$\max_{x_1} \{g_1(x_1 - x_0) + \max_{x_2, x_3, ..., x_N} [g_2(x_2 - x_1) + g_3(x_3 - x_2) + g_N(x_N - x_{N-1})]\}.$$

This suggests the way in which the dynamic programming functions can be identified. We define $f_k(y)$ as the optimal return obtainable from the $(N-k)$ functions $g_{k+1}, ..., g_N$, where $y$ is the value of $x_{k-1}$. Thus

$$f_k(y) = \max_{x_k, x_{k+1}, ..., x_N} \{g_k(x_k - y) + g_{k+1}(x_{k+1} - x_k) + ... + g_N(x_N - x_{N-1})\}.$$

Then we can obtain the solutions by the recurrence relations

$$(f_k y) = \max_{x_k \geq 0} \{g_k(x_k - y) + f_{k+1}(x_k)\}$$

which are evaluated for an appropriate set of $x_k$ values and for $k = N, N-1, N-2, ..., 1$ in that order. The values of $y$ are the permissible values of $x_{k-1}$. Again it is comparatively easy to see how the recurrence relations can be solved once the problem has been formulated.

## *7.11 A stochastic problem

Finally we will illustrate how dynamic programming procedures can readily treat problems with probabilistic features by examining a simple problem rather strangely called the 'Flyaway-Kit Problem'. Suppose there are $N$ items held in a depot, which are in demand, and that the demand for each item is distributed in a known fashion. Let $p_i(j)$ be the probability of a demand for $j$ items of type $i$ occurring. There is a capacity restriction $S$ on the number of items which may be held at the depot, and the volume of item $i$ is $s_i$. The cost of being 1 unit short on item $i$ is $c_i$. We wish to determine how to load the capacity $S$ so as to minimize the expected cost.

Let $x_i$ be the number of items of type $i$ which are loaded. Then the expected cost due to unfulfilled demand is:

$$c_i \sum_{j=x_i+1}^{\infty} p_i(j)(j-x_i).$$

The total expected cost is

$$F(x_1, x_2, ..., x_N) = \sum_{i=1}^{N} c_i \left[ \sum_{j=x_i+1}^{\infty} p_i(j)(j-x_i) \right].$$

The problem is to minimize $F$ subject to

$$\sum_{i=1}^{N} x_i s_i \leq S$$

and

$x_i = 0, 1, 2, ...,$ for all $i$.

The previous examples suggest how this problem can be tackled. Define $f_k(y)$ as the cost associated with an optimal choice of items of the first $k$ types if a total capacity of $y$ has been used where $0 \leq y \leq S$. Then the basic recurrence relation is

$$f_k(y) = \min_{0 \leq x_k \leq \left[\frac{y}{s_k}\right]} \left[ c_k \sum_{j=x_k+1}^{\infty} p_k(j)(j-x_k) + f_{k-1}(y-x_k s_k) \right].$$

The solution to this problem is essentially similar to the deterministic allocation problems which have already been discussed.

REFERENCES

Bellman, R. E. and Dreyfus, S. E. 1962. *Applied Dynamic Programming*. Princeton.
Hadley, G. 1964. *Non-linear and Dynamic Programming*. Addison-Wesley, New York.
Howard, R. A. 1966. Dynamic programming. *Man. Sci.*, **12**, 317.
Nemhauser, G. L. 1966. *Introduction to Dynamic Programming*. Wiley, New York.

## Exercises on Chapter 7

**1** Determine the shortest route from point 1 to point 10 in the following network using the dynamic programming procedure. The distances between the points are marked in the diagram.

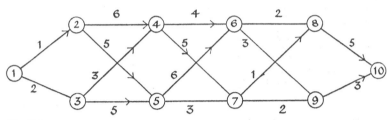

Fig. 7.4

**2** A resource can be added in total or in part to any of 4 productive operations. An allocation of $x$ to operation $i$ gives a profit of $g_i(x)$. If the total amount available is 6 and only integral allocations are permitted, find the optimal allocations where the $g_i(x)$ are given as shown in Table 7.6.

**Table 7.6**

| $x$ | $g_1(x)$ | $g_2(x)$ | $g_3(x)$ | $g_4(x)$ |
|-----|----------|----------|----------|----------|
| 0 | 2 | 0 | 1 | 0 |
| 1 | 4 | 1 | 3 | 3 |
| 2 | 6 | 2 | 5 | 6 |
| 3 | 7 | 3 | 7 | 6 |
| 4 | 7 | 4 | 8 | 6 |
| 5 | 7 | 6 | 8 | 9 |
| 6 | 7 | 8 | 9 | 12 |

Supposing only 3 units were available, what would be the optimal allocation pattern?

**3** In a cargo-loading problem we are given a set of 4 items where the weights and values per item are as shown in Table 7.7. Determine the numbers of

**Table 7.7**

| Item | 1 | 2 | 3 | 4 |
|------|---|---|---|----|
| Weight | 1 | 3 | 4 | 6 |
| Value | 1 | 5 | 7 | 11 |

each item to be loaded so as to maximize the value of the cargo subject to not exceeding a total weight of 11.

**4** A machine is composed of 3 different types of component in series and each type of component may be duplicated in parallel one or more times to increase reliability. If $x_i$ is the number of components of the $k$th type which are in the system, the probability of successful operation at the $k$th stage is $1-(\frac{1}{2})^{x_k}$. If the cost of the 3 different types of components are 1, 2, 3 respectively determine how many components should be placed in each stage so as to maximize the reliability of the system if the total cost permitted is 9.

**5** Show how dynamic programming can be used to solve the maximization problem:

Maximize $\sum\limits_{k=1}^{N} g_k(x_k)$

subject to $x_1, x_2, x_3, \ldots, x_N = C.$

$\qquad x_i \geq 1.$

Notice that in reliability terms this problem may be interpreted as the minimization of a cost subject to attaining a given reliability, rather than maximizing the reliability for a given cost.

**6** In some applications we wish to minimize a maximum deviation subject to a limitation on the average deviation. This type of problem can be expressed as

Minimize $[\max (g_1(x_1), g_2(x_2), \ldots, g_N(x_N))]$

subject to $x_1 + x_2 + \ldots + x_N = C.$

Derive the dynamic programming recurrence relations for this problem.

**7** Use calculus to solve the recurrence equations in the dynamic programming solution of the problem:

Minimize $x_1^2 + x_2^2 + x_3^2$

subject to $x_1 + x_2 + x_3 = C$

$\qquad x_i \geq 0, \quad i = 1, 2, 3.$

**\*8** In general a network is not structured so conveniently as Example 7.1. A general network is specified as a set of $N$ points with connections between pairs of points. Let $d(i, j)$ denote the distance from point $i$ to point $j$. (Assume $d(i, j) = \infty$ if the points are not directly connected.) Construct a dynamic programming procedure for finding the shortest route from point 1 to point $N$.

**\*9** The travelling salesman problem is to find a minimum tour through $N$ cities, starting from a given city, going through each city once and returning to the starting point. The distance between each pair of cities is known. Attempt to formulate this problem in terms of dynamic programming and point out the difficulties which occur.

**\*10** What briefly are the consequences of not assuming monotonicity in the $g_k(x_k)$ functions of the resource allocation problem of Section 7.6?

**\*11** The ideas of dynamic programming can also be used to solve puzzles. Suppose a mixture requires that 7 grams of a material be measured out, but the measuring instruments consist simply of a 5-gram container and an 8-gram container. Assuming both containers are initially empty and that there is a large bag of the material to draw on, how can the 7 grams be determined? Consider the problem as a multi-stage decision process starting from the final state required.

# 8 Branch and bound methods

## 8.1 Solutions as tree searches

The branch and bound method is one of the most recently developed optimization techniques and it has achieved a notable success on a certain class of problems. It is particularly suited to well-structured problems with integer constraints on the variables. Like dynamic programming it does not deal with

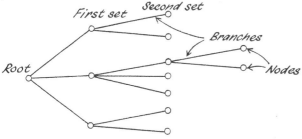

Fig. 8.1

a specific mathematical framework nor does it follow the conventional iterative idea of an optimization process. Its aim is to conduct a reduced search over all possible solutions, the reduction being dependent on how well the problem structure can be exploited.

The basic idea is to picture the construction of a solution to an optimization problem as a search over a decision tree. The tree consists of branches, branch points called nodes and a root for the tree as illustrated in Fig. 8.1.

123

The root of the tree out of which all the branches emerge (the node on the extreme left in Fig. 8.1 corresponds to all possible solutions of the problem. The branches coming out of the root correspond to various subdivisions of the problem and the subsequent branches out of other nodes correspond to further subdivisions. At each subdivision point or node, a decision is made which is usually the determination of the value of a variable. For example if we determine the value of the variable $x_1$ first and there are three possible values for $x_1$ we create three branches leading to three nodes; then we have three new sub-problems in $(N-1)$ variables, if there were a total of $N$ variables, as whichever node we branch from a value for $x_1$ is fixed. The next set of nodes may correspond to a further decision about the variable $x_2$ given the particular value of $x_1$ from which the branch stems. Thus the nodes correspond to incomplete or partial solutions at which some of the variables have values and some do not, and the branches leading up to the node specify the partial solution.

The total tree corresponds to all possible solutions of the problem, but the aim of the branch and bound method is to cut down dramatically on the total search. By estimating what are the best possible results which could be obtained down a branch early on in the construction of the branch, we hope to be able to eliminate masses of possible solutions. These estimates are called the bounds. It is only if this estimation and elimination procedure works well that the branch and bound idea of subdividing the problem can be successful.

The branch and bound method can readily handle constraints. We are always in control of the development of a solution in the course of constructing the alternatives down the branches so that the constraints can have any form. For example the method can readily handle the 'either-or' type of restriction mentioned in Section 3.8, and we have not yet considered a method capable of this.

We will now consider a small example to illustrate the technique, and then we will characterize the general rules of the method. Suppose we have a problem in four variables $x_1$, $x_2$, $x_3$, $x_4$ each of which may assume the values 1 or 2 and an objective function $F(x_1, x_2, x_3, x_4)$ which is to be minimized. We can construct the optimal solution by first assigning a value to $x_1$, then to $x_2$, then to $x_3$ and $x_4$ in turn. Therefore the first decision is to consider whether $x_1$ should take the value 1 or 2. This will subdivide the problem into two parts. We may represent this subdivision in the tree structure form as shown in Fig. 8.2.

We now want to choose either of the branches node 1 to node 2 or node 1 to node 3 and extend the branch to consider the values $x_2 = 1$ or 2. To choose which node to branch from we require an estimate of what is the least value which the objective function may take down either of these branches. Assuming this can be estimated, we associate a lower bound with node 2 as $B(2)$ and a lower bound with node 3 as $B(3)$. Suppose $B(2) < B(3)$, it would then seem to be best to branch out from $B(2)$ as the problem is a minimization.

We would thus subdivide the solution along this branch by constructing two new nodes 4 and 5 with new bounds $B(4)$ and $B(5)$ associated with the partial solutions $x_1 = 1$, $x_2 = 1$, and $x_1 = 1$, $x_2 = 2$. It should be noted that these bounds will be at least as large as $B(2)$. The tree structure would expand as shown in Fig. 8.3.

The next step would be to branch out from nodes 3, 4 or 5 (as these form

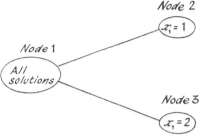

Fig. 8.2

the front of incomplete solutions nodes) and again the best choice would be to branch from the node with the smallest lower bound.

By continuing the branching and bounding process, we would ultimately obtain a complete solution $(\bar{x}_1, \bar{x}_2, \bar{x}_3, \bar{x}_4)$ say at node $\bar{i}$ with a bound $B(\bar{i})$ in which values have been assigned to all variables. We would then check to see if there were any bounds $B(i)$ at the end of branches for which $B(i) < B(\bar{i})$. If there were none this solution would be optimal. If there were, we would

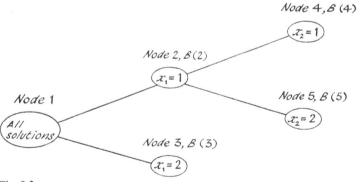

Fig. 8.3

have to continue the procedure until we had a complete solution such that the continuation of any branch could not offer a lower objective function value.

## 8.2 Rules for the branch and bound method

The previous section has illustrated the method and the associated tree structure of nodes which is used to present the construction of the solution. Although the method is very adaptable, and the descriptions of the method vary widely, there are four main parts to the specification of a branch and bound procedure: the node structure, the branching policy, the lower bound calculations and a terminating rule.

### The tree structure

The first task is to specify the tree structure of the solution. This means identifying what the nodes will represent. The initial node is 'all solutions'. The next set of nodes might be all possible values for a variable, or they might represent the assignment of least values to each of the variables. We therefore specify what the successive nodes will stand for, and, associated with this, how many branches will emerge from each node.

### The branching policy

The branching policy decides which node we choose to branch from next. In the case of a minimization problem we will usually branch from the node which currently has the least lower bound value. (For a maximization we choose the greatest upper bound.) Another policy is to branch from the best node of the last set of nodes to be created. This has the advantage of reaching a complete solution quickly even if more nodes have to be evaluated in the end.

### Formulae for the bounds

A formula must be specified for calculating lower bounds (upper bounds in the case of a maximization problem). This is usually the most difficult task. It requires detailed insight into the nature of the problem. Efficient bounding formulae will terminate unpromising branches quickly. But the existence of such formulae requires a good deal of problem structure which can be exploited. There is great scope for ingenuity in each particular case.

### A terminating rule

Fourthly, we need a means of recognizing when the optimal solution has been obtained. In the case of a minimization problem this is recognized by the fact that a complete solution has been obtained at node $M$ whose objective function value $B(M)$ is less than or equal to any other bound $B(r)$ where $r$ is a node at the end of any branch. Although this optimal termination is obvious, it is important to include a terminating rule explicitly in the procedure as often we may wish to stop the calculation short of obtaining the global optimum. These suboptimization possibilities are discussed in a later section.

We will now give two examples of problems amenable to the branch and bound treatment. As an advance warning, some of the working in these examples is complex because of the ingenuity required in the estimation of lower bounds. The study of worked cases is the only way in which the potential and the difficulties of the method can be recognized. Subsequently we will see how branch and bound can be usefully applied to integer-linear programming problems.

## 8.3 The three-machine scheduling problem

The three-machine scheduling problem may be stated as follows. Each of $N$ jobs has to be processed through each of three machines $A$, $B$, $C$ in the same order. The time of job $j$ on the machines is $a_j$, $b_j$, $c_j$ respectively. A machine can only process one job at a time, and two operations of one job cannot be done concurrently. The problem is to schedule the jobs through the machines so as to minimize the total production time required. As, with only three machines, there is no advantage from overtaking, the problem is to order the jobs on machine $A$ and the objective is equivalent to minimizing the time at which the last job is finished on machine $C$.

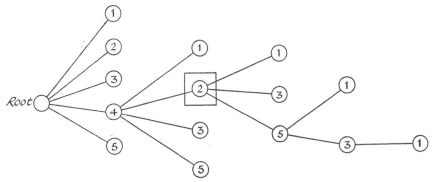

Fig. 8.4

A branch and bound procedure can now be prescribed. First we decide on the tree structure. As usual all solutions start from the root of the tree. The first node out of the root will be the first job to be processed. If there are $N$ jobs we require $N$ branches out of the root to cover all possibilities for first place in the sequence. Associated with any first node there will be $(N-1)$ jobs which could be placed in the second position so there will be $(N-1)$ possible branches out of the first set of nodes. By the same argument there will be $(N-2)$ jobs and $(N-2)$ branches for a third node leading down a branch from the root. For example, if there were 5 jobs there would be 5 branches out of the root, 4 branches out of a node one link away from the root, 3 branches out of any of these nodes, etc. Fig. 8.4 shows a sequence of nodes corresponding to the job sequence 4, 2, 5, 3, 1. Any particular node defines a subsequence of jobs which have been scheduled in a definite order and a set of jobs which have still to be scheduled. For example, the node in Fig. 8.4 with the square around it corresponds to the sequence of two jobs

4, 2 having been scheduled and the ordering of the jobs 5, 3 and 1 not yet decided.

As a branching policy we will branch next from the node with the least lower bound, and we will branch out to $n$ nodes if there are $n$ unscheduled jobs.

It remains to determine a lower bound on the completion time on machine $C$ when a subsequence of jobs has been scheduled and some jobs remain still to be sequenced. The first piece of information to note towards calculating the lower bound is the times at which the machines become empty after the given subsequence has been scheduled. These machine empty times are best calculated iteratively, by updating their values each time a job is added to the subsequence. Suppose in Fig. 8.4 we wanted to calculate the lower bounds for the sequence 4, 2, 5 and that the machine empty times resulting from the subsequence 4, 2 were $T_a$ (4, 2), $T_b$ (4, 2), $T_c$ (4, 2) as shown in Fig. 8.5. We now want to calculate the times $T_a$ (4, 2, 5), $T_b$ (4, 2, 5), $T_c$ (4, 2, 5) at which the machines become empty if job 5 is placed on the end of the sequence.

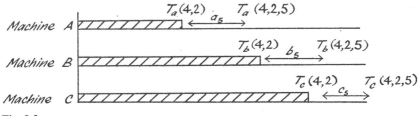

Fig. 8.5

The value of $T_a$ (4, 2, 5) will be simply $T_a$ (4, 2) plus the duration of job 5, i.e.

$T_a$ (4, 2, 5) = $T_a$ (4, 2)$+a_5$.

The value of $T_b$ (4, 2, 5) will be the time at which job 5 starts on machine $B$ plus the duration of job 5. Job 5 starts on machine $B$ either at the time it finishes on machine $A$ or the time at which machine $B$ becomes empty whichever is the larger number. (In the diagram $T_b$ (4, 2)$>T_a$ (4, 2, 5).) Therefore $T_b$ (4, 2, 5) is calculated as

$T_b$ (4, 2, 5) = max ($T_a$ (4, 2, 5), $T_b$ (4, 2))$+b_5$.

Similarly job 5 starts on machine $C$ at the time it finishes on machine $B$, $T_b$ (4, 2, 5) or the time machine $C$ becomes empty $T_c$ (4, 2) whichever is the larger number. (In the diagram $T_c$ (4, 2)$<T_b$ (4, 2, 5).) Therefore

$T_c$ (4, 2, 5) = max ($T_b$ (4, 2, 5), $T_c$ (4, 2))$+c_5$.

We have thus calculated the times at which the machines become empty for the sequence 4, 2, 5 and we now have to estimate how much time will be required on machine $C$ before the last job is complete as we still have to schedule jobs 1 and 3 in some order. The estimate really depends on whether machines $A$ or $B$ or $C$ is going to be the bottleneck machine consuming the most time. Suppose machine $A$ is the bottleneck. Then we require at least a time $a_1 + a_3$ on machine $A$ after time $T_a$ (4, 2, 5) plus the least time it takes to get the last of the unscheduled jobs through machines $B$ and $C$, i.e.

$$\min_{j=1,3} (b_j + c_j).$$

Therefore a lower bound $L_a$ on the finish time on machine $C$ with machine $A$ as the bottleneck is

$$L_a \; (4, 2, 5) = T_a \; (4, 2, 5) + a_1 + a_3 + \min (b_1 + c_1, b_3 + c_3).$$

On the other hand, machine $B$ may be the bottleneck. In this case we need a time of at least $T_b$ (4, 2, 5) $+ b_1 + b_3$ on machine $B$ plus the time it takes to get the last job through machine $C$. Thus a lower bound $L_b$ (4, 2, 5) on the finish time on machine $C$ if machine $B$ is the bottleneck is

$$L_b \; (4, 2, 5) = T_b \; (4, 2, 5) + b_1 + b_3 + \min (c_1, c_3).$$

Finally, if machine $C$ is the bottleneck the least time it takes to complete all the jobs on machine $C$ is $T_c$ (4, 2, 5) plus the time it takes to process the remaining jobs on machine $C$, giving the lower bound $L_c$ (4, 2, 5) as

$$L_c \; (4, 2, 5) = T_c \; (4, 2, 5) + c_1 + c_3.$$

We have therefore calculated three lower bounds on the completion time on machine $C$. These bounds are all at least as small as the actual completion time which can be achieved, and therefore we take the largest of the lower bounds as the estimate of the completion time. The final lower bound $LB$ (4, 2, 5) is therefore

$$LB \; (4, 2, 5) = \max (L_a \; (4, 2, 5), L_b \; (4, 2, 5), L_c \; (4, 2, 5)).$$

The formula for the lower bounds can now be written for the general case. Suppose we are at a node $n$ at which a subsequence of jobs $S_n$ has been scheduled and a subset $E_n$ have still to be scheduled. Then the lower bound associated with the subsequence $S_n$ denoted by $LB(n)$ is

$$LB(n) = \max \begin{cases} T_a(S_n) + \sum_j a_j + \min_j (b_j + c_j) \\ T_b(S_n) + \sum_j b_j + \min_j c_j \\ T_c(S_n) + \sum_j c_j \end{cases}$$

where the summations and minimizations extend over all values of $j$ for which $j \in E_n$.

*Example 8.1*

We can now apply this procedure to the following example with four jobs in which the time the various jobs take on the machine are given in Table 8.1. The solution is expressed in the tree drawn out in Fig. 8.6 with 17 nodes. The lower bounds $LB(r)$ for node $r$ are marked and are numbered according to their order of calculation. The numbers in the nodes are the jobs in the various positions in the sequence. For instance, the top branch corresponds to the sequence 1, 2, 3. It should be noted that after three jobs have been

**Table 8.1**

| Job $j$ | $a_j$ | $b_j$ | $c_j$ |
|---------|-------|-------|-------|
| 1 | 13 | 3 | 12 |
| 2 | 7 | 12 | 16 |
| 3 | 26 | 9 | 7 |
| 4 | 2 | 6 | 1 |

placed in the sequence the lower bound is equivalent to the total production time if the final job is placed on the end of the branch, so that the tree is not extended beyond the third set of nodes.

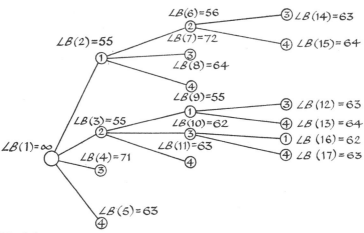

Fig. 8.6

The lower bounds are calculated by the formulae described. For example the lower bound $LB(9) = 55$ is calculated as follows. Denoting the sequence 2, 1 by $J_9$, we see from Fig. 8.7 that $A(J_9) = 20$, $B(J_9) = 23$ and $C(J_9) = 47$.

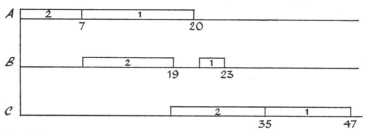

Fig. 8.7

Also

$$\sum_{i=3,4} a_i + \min_{i=3,4} (b_i + c_i) = 28 + \min (16, 7) = 35$$

$$\sum_{i=3,4} b_i + \min_{i=3,4} c_i = 15 + \min (7, 1) = 16$$

$$\sum_{i=3,4} c_i = 8.$$

Therefore

$$LB(9) = \max (55, 39, 55) = 55.$$

It can be seen that the lower bound based on machine $B$ would give a very poor estimate of the completion time. The optimal sequence is expressed in the branch leading to node 16 where the total production time (finish time on machine $C$) is 62 units.

## 8.4 The knapsack problem

The knapsack problem is very similar to the cargo-loading problem discussed in the last chapter, but in this case we have only one item of each type to consider. The problem is to determine the maximum value collection of items which do not exceed a limited weight. Let there be $N$ items, let $w$ be the maximum weight allowed, and let $v(i)$ and $w(i)$ denote the value and weight of the $i$th item. The problem is to choose the set $S$ of items such that

$$\sum_{i \in S} v(i) \text{ is a maximum}$$

and

$$\sum_{i \in S} w(i) \leq W.$$

We will adopt the following branch and bound policy. First we will order the items by decreasing value per unit weight ratio. If the indices of the items in this order are $i_1, i_2, i_3, \ldots, i_N$, the item $i_1$ is the most valuable per unit weight, $i_2$ the next most valuable per unit weight and so on. Now, starting from all possible solutions we will subdivide the problem into two branches either containing item $i_i$ or not containing item $i_1$. We will mark these nodes as $i_1$ and $i_1^*$. We will then branch from one of these nodes subdividing the solution into two further nodes so that it contains $i_2$ or not and marked $i_2$ and $i_2^*$. We thus obtain a tree of the form shown in Fig. 8.8.

Any given branch will determine the inclusion of certain items and the exclusion of others. Suppose at node $r$ there is a set $E_r$ of excluded items and a set $I_r$ of included items. Then we wish to calculate the upper bound $UB(r)$ (as it is a maximization problem) associated with node $r$. This is where the ingenuity is needed. We imagine that it is possible to split the items into parts and we know we cannot do better than select the best value for weight ratios assuming items are splittable. For example if we had three items numbered 1, 2, 3 with values 7, 4, 1, and weights 3, 2, 1, respectively, the best way to fill a weight capacity of 4 would be to include items 1 and 3 giving a value of 8 and using the full capacity. However, if we assume the items are splittable we would take the whole of item 1 with the best value/weight ratio and half of item 2 with the next best value/weight ratio, thus using up the full capacity of 4 and giving the improved value of $7 + \frac{1}{2}.4 = 9$. We therefore calculate the upper bound $UB(r)$ for node $r$ as:

$$UB(r) = \sum_{j \in I_r} v(j) + \sum_{\substack{j = 1 \\ ij \in E_r}}^{k} v(i_j) + f \cdot v(i_{k+1})$$

where the index $k$ and the fraction $f$ are chosen so that

$$\sum_{i \in I_r} w(i) + \sum_{\substack{j = 1 \\ ij \in E_r}}^{k} w(i_j) + f \cdot w(i_{k+1}) = W.$$

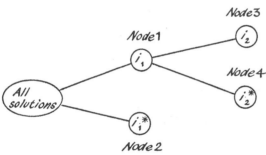

Fig. 8.8

Sometimes at a node the weight constraint may be violated and then we will allocate a value 0 to $UB(r)$. It is violated if $\sum_{i \in I_r} w(i) > W$. We will branch from the node with the highest upper bound, and we continue branching until all items in a branch belong to $I_r$ or $E_r$ and there is no node with a larger upper bound which has not been extended.

*Example 8.2*

We will now apply the procedure to the following example of 7 items where $W = 100$, given in Table 8.2.

We now compute the ratios $v(i)/w(i)$ and re-order the items as in Table 8.3. The sequence of nodes and bound values $UB(r)$ are then obtained as shown in the tree of Fig. 8.9. The numbers in the nodes correspond to the new indices of the items.

The upper bound calculations are obtained by the formula expressed above. For example, the upper bound at node 6, $UB(6)$, contains items with new indices 1 and 2 and not 3. This leads to a total value of items 1, 2, 4, and one-quarter of five as at this point the weight restriction is exactly satisfied.

**Table 8.2**

| Item No. | Weight | Value |
|:--------:|:------:|:-----:|
| 1 | 40 | 40 |
| 2 | 50 | 60 |
| 3 | 30 | 10 |
| 4 | 10 | 10 |
| 5 | 10 | 3 |
| 6 | 40 | 20 |
| 7 | 30 | 60 |

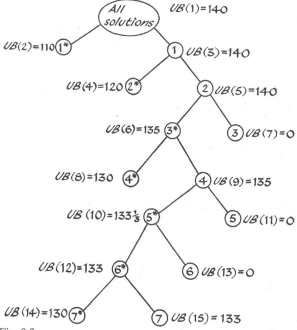

Fig. 8.9

**Table 8.3**

| New Index | Item No. | Weight | Value | Ratio |
|:---------:|:--------:|:------:|:-----:|:-----:|
| 1 | 7 | 30 | 60 | $2$ |
| 2 | 2 | 50 | 60 | $\frac{6}{5}$ |
| 3 | 1 | 40 | 40 | $1$ |
| 4 | 4 | 10 | 10 | $1$ |
| 5 | 6 | 40 | 20 | $\frac{1}{2}$ |
| 6 | 3 | 30 | 10 | $\frac{1}{3}$ |
| 7 | 5 | 10 | 3 | $\frac{3}{10}$ |

Therefore

$$UB(6) = 60 + 60 + 10 + \tfrac{1}{4}.20 = 135.$$

Note also that $UB(7) = 0$ as the sum of the weights of items 1, 2 and 3 is 120 which exceeds the weight limit of 100.

The optimum result obtained at node 15 is shown to contain items 1, 2, 4 and 7 and to have a value of 133.

## 8.5 Integer programming

The branch and bound method offers a valuable extension to the linear and non-linear programming methods when some or all of the variables are constrained to be integer valued. Many optimization problems arise in which some of the variables are constrained to be integer valued such as nominal pipe size, transformer lamination thickness, number of distillation column plates, or number of personnel. The general form of an integer-programming problem in $N$ variables in which $n$ variables must be integer is:

Minimize $F(x_1, x_2, ..., x_N)$

subject to $G_k(x_1, x_2, ..., x_N) \leq 0$ for $k = 1, ..., M$

and $x_1, x_2, ..., x_n$, integral, $n \leq N$.

Usually the $G_k$ constraints will contain bounds on the integer valued variables such as

$$b_j \leq x_j \leq B_j$$

which restrict $x_j$ to a few integer values such as 2, 3 or 4, an important case being the zero-one constraint that

$$x_j = 0 \text{ or } 1.$$

If the variable $x_j$ can assume a wide range of values, say 0 to 50, then it is usually reasonable to allow $x_j$ to vary continuously in the interval and round the value produced by the continuous variable optimization methods to the nearest integer value. Therefore we are primarily concerned with situations in which the variable must take on a few values.

We will consider the context of integer linear programming and the scheme which will be proposed readily extends to the non-linear case. The problem is to

minimize $\displaystyle\sum_{i=1}^{N} c_i x_i$

subject to $\displaystyle\sum_{j=1}^{N} a_{ij} x_j \leq b_i$, for $i = 1, ..., M$

$$0 \leq x_j \leq B_j, \ x_j \text{ integral for } j = 1, ..., k$$

and $x_j \geq 0$ for $j = k+1, N$.

The integers $B_j$ are assumed to be comparatively small numbers. (If the lower limits on the integer variables is not 0 but $-d_j$, say, we can rewrite the problem with $y_j + d_j$ replacing $x_j$ throughout giving the limits on $y_j$, as 0 and $B_j + d_j$.) The branch and bound method is combined with linear programming to solve this problem as follows. First we solve the linear programming problem ignoring the integer constraints. This gives the lower bound associated with the first node (corresponding to all possible solutions) with lower bound say $B(0)$. If all $x_j$, $j = 1, \ldots, k$ happen to be integer valued then the solution is optimal. For clearly the minimum to an unconstrained problem is at least as small as a minimum with added constraints. Suppose at least one of the integer variables $x_j$ is not integer valued, and let it be $x_1$. Then we branch to create $(B_1 + 1)$ nodes corresponding to the values $x_1 = 0, 1, 2, \ldots, B_1$ and re-solve the $(N-1)$ variable linear programming problem $B_1 + 1$ times with $x_1$

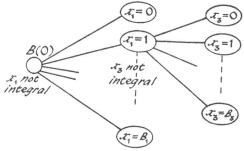

Fig. 8.10

set to each of these values to provide lower bounds. Now say the least of these values occurs at $\bar{x}_1$, we can inspect to see whether any of the other integer variables are not integer valued at this solution, and repeat the whole process with the next variable. Fig. 8.10 illustrates the process where $\bar{x}_1 = 1$ and the next branching variable is $x_3$.

If the objective function is to be maximized, the same scheme is used but the bounds will be upper bounds rather than lower bounds.

*Example 8.3*

Solve the integer linear programming problem:

Maximize $x_1 + x_2$

subject to $5x_1 + 3x_2 \leq 15$

$\qquad 4x_1 + 5x_2 \leq 20$

$\qquad x_1, x_2 \geq 0$ and integral.

From Fig. 8.11 it is clear that $x_1 \leq 3$ and $x_2 \leq 4$. The maximum value of the objective function is $4\frac{3}{13}$, occurring at the point $(1\frac{2}{13}, 3\frac{1}{13})$. This is the first upper bound. We now try the integer solutions $x_1 = 0, 1, 2, 3$ and obtain maximum objective function values for the corresponding linear programming problems as 4, $4\frac{1}{5}$, $3\frac{2}{3}$, 3. Taking $x_1 = 1$, for which $x_2 = 3\frac{1}{5}$, we evaluate $x_2 = 0, 1, 2, 3, 4$ and obtain objective function values 1, 2, 3, 4. The value $x_2 = 4$ is not feasible. Since the objective function for $x_2 = 3$ is as large as any bound obtained at previous nodes this is the optimum solution. The tree

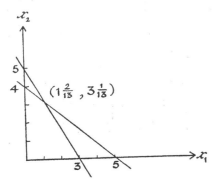

Fig. 8.11

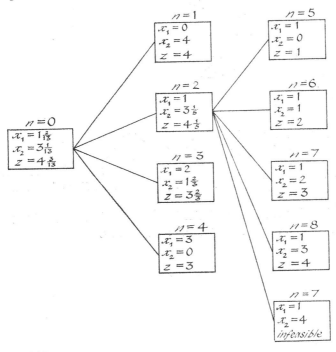

Fig. 8.12

of nodes is shown in Fig. 8.12 where the objective function value is written as $z$ and the order of evaluating the nodes is given by the index $n$.

It may correctly be commented that the proposed branching policy is weak because we can expect the integer optimum to occur somewhere near the original non-integer optimum. In Example 8.3 just considered, the integer optimum solution was at (1, 3) which is very close to the initial optimum solution $(1\frac{2}{13}, 3\frac{1}{13})$. In general, if $\bar{x}_j$ is the non-integer value for the integer constrained variable $x_j$ in a current solution we could try branching to the values

$$x_j = [\bar{x}_j]$$

and

$$x_j = [\bar{x}_j] + 1$$

(where $[x]$ is the largest integer less than $x$) to create the two integer values on either side of $\bar{x}_j$. In fact the bounds will decrease for other integer values

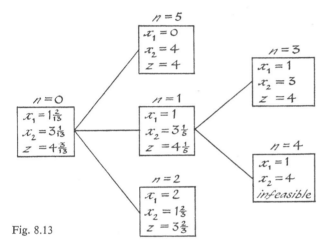

Fig. 8.13

of $x_j$ on either side of the values $[\bar{x}_j]$ and $([\bar{x}_j] + 1)$ as the feasible objective function values constrained by the previously fixed integer values will be concave in $x_j$ (convex for a minimization problem). We can take advantage of this feature to examine only two branches out of a node at the first extension but we may well have to back-track to the node to try other possibilities later in the search.

If this scheme is applied to Example 8.3 we would evaluate the solutions in the order indicated in Fig. 8.13 by the index $n$. After obtaining solutions at nodes $n = 1$ and $n = 2$ with $x_1$ values of 1 and 2 which flank the best value $x_1 = 1\frac{2}{13}$, we press on to $n = 3$ and $n = 4$ with $x_2 = 3$ and $x_2 = 4$ which flank $x_2 = 3\frac{1}{5}$. The best objective function value so far is $z = 4$ for $n = 3$. But it is now conceivable that we could obtain a solution giving a $z$ value greater than 4 and just less than $4\frac{1}{3}$ for $x_1 = 0$. We therefore evaluate the solution at node $n = 5$ giving $z = 4$ which is equivalent to the best solution obtained. It should be noticed that this is the only branch and bound scheme in which we may have to backtrack to a node from which we have already branched.

## 8.6 Computational considerations

The branch and bound method is unfortunately of limited application. Quite apart from the difficulty of calculating bounds the amount of computation which is necessary to solve problems increases exponentially as the problems get larger. Also the amount of computation which is necessary for any given size of problem varies with the particular data under consideration.

In the three-machine scheduling problem of Section 8.3, if there were $N$ jobs the minimum number of nodes which would need to be evaluated would be

$$N + (N-1) + (N-2) + \ldots + 2 + 1 = \frac{N(N+1)}{2}.$$

This seems quite reasonable. But on the other hand the maximum number would be

$$N + N(N-1) + N(N-1)(N-2) + \ldots N! \simeq e.N!$$

which is a very large number. Equally on the knapsack problem, with $N$ items the minimum number of nodes is $2 + 2 + \ldots + 2 \ldots + 2 = 2N$, whereas the maximum number is $2 \times 2 \times 2 \ldots \times 2 = 2^N$. The difference between these limits is very large and we cannot tell in advance how much calculation will be required. The upper limits grow rapidly with increasing values of $N$.

## 8.7 Suboptimization

One policy which has sometimes been adopted to overcome the computational hazards is to aim at a suboptimal result which is within a certain percentage of the optimum. The branch and bound procedure is capable of handling this type of request, because we can continue down the branch which is nearest to a complete solution, and only require that the lower bounds of the incomplete branches should be within, say, 5 or 10 per cent of the final objective function value obtained. Although this opportunity extends the scope of the branch and bound method to larger problems, where some degree of suboptimization can be tolerated, the number of variables will usually remain comparatively small on the average problem, say 25 at the most, excepting some particularly agreeable problems where a very efficient bounding scheme is available.

REFERENCES

Agin, N. 1966. Optimum seeking with branch and bound. *Man. Sci.*, **13** (4), 176.
Balas, E. 1965. An additive algorithm for solving linear programs with zero-one variables *Opns. Res.*, **13** (4), 517.

Balinski, H. L. 1965. Integer programming: methods, uses, computation. *Man. Sci.*, **12** (3), 325.

Ignall, E. and Shrage, L. 1965. Application of the branch and bound method to some flow-shop scheduling problems. *Opns. Res.*, **13** (3), 400.

Kolesar, P. J. 1967. A branch and bound algorithm for the knapsack problem. *Man. Sci.*, **13** (9), 723.

Lawler, E. and Wood, D. E. 1966. Branch and bound methods: a summary. *Opns. Res.*, **14** (4), 699.

Little, J. D. C., Murty, K. G., Sweeney, D. W. and Karel, C. 1963. An algorithm for the travelling salesman problem. *Opns. Res.*, **11**, 972.

## Exercises on Chapter 8

**1** Devise a branch and bound method for the two-machine scheduling problem similar to the three-machine problem described in Section 8.3. Assume the jobs go through both machines in the same order. Use it to determine the job sequence which minimizes the time of completion of the last job on the following data.

**Table 8.4**

| Job $j$ | $a_j$ | $b_j$ |
|---------|-------|-------|
| 1 | 4 | 1 |
| 2 | 2 | 6 |
| 3 | 4 | 5 |
| 4 | 1 | 2 |

**2** Apply the branch and bound procedure to the following five-item knapsack problem in which the maximum weight capacity is 9. (The items are already ordered by value/weight ratio.)

**Table 8.5**

| Item index | Value | Weight |
|------------|-------|--------|
| 1 | 10 | 4 |
| 2 | 6 | 3 |
| 3 | 12 | 6 |
| 4 | 3 | 2 |
| 5 | 2 | 2 |

**3** Use the branch and bound method to solve the integer linear programming problem:

Maximize $5x_1 + 2x_2$

subject to $5x_1 + x_2 \leq 17$

$$x_1 + 5x_2 \leq 25$$

$$x_1, x_2 \geq 0 \text{ and integral.}$$

**\*4** Devise a branch and bound procedure for the travelling salesman problem where the task is to find the shortest route passing through each of $N$ cities precisely once, where $d(i, j)$ is the distance from city $i$ to city $j$.

**\*5** In the three-machine scheduling problem described in Section 8.3 it was assumed that the jobs went through the three machines in the same order. Investigate how the branch and bound scheme would have to be changed if the jobs went through the three machines in various orders, not all $A$ to $B$ to $C$, where the objective is to minimize the time at which the last job is completed.

# 9 Permutation procedures

## 9.1 The need to suboptimize

Despite the capability of the range of techniques which have been described there are still plenty of optimization problems which they cannot handle. The problem may be too large to solve in a reasonable amount of computer time by a formal technique or the structure may be too awkward to be accommodated within the mathematical frameworks which have been proposed. Often a cruder approach must be adopted in which we never aim to reach the global optimum but set out simply to obtain a good answer in as systematic a fashion as possible. These next two chapters on techniques will examine suboptimal methods of this type. Here we will study a method for optimizing large permutation problems, and in the next chapter we will describe the main characteristics of heuristic techniques suitable for allocation problems.

## 9.2 Permutation problems

We have already encountered a number of problems which express themselves as permutations. The travelling salesman problem and the three-machine scheduling problem are essentially problems of finding the best sequence of a number of elements and we have investigated their solution by dynamic programming and branch and bound procedures. However, these methods will only solve comparatively small cases and in practice we may have to tackle problems of ordering a large number of elements. There are a variety of important industrial problems of this type. For instance in the ordering of operations through a machine centre, the placement of logic elements on a circuit board, or the route which a delivery vehicle should take through a number of demand points are all permutation problems. A variety of congestion problems can also be represented as an ordered sequence or permutation. Before investigating how we can cope with these large permutation problems, we will define the general problem of optimizing a permutation.

A permutation consists of an ordering of a number of elements. Let there be $N$ elements labelled $e_1, e_2, ..., e_N$. Let $p_j$ denote the element occupying position $j$ in the sequence, and denote the complete permutation by $[P]$ where

$$[P] = [p_1, p_2, ..., p_N].$$

(The square brackets are used to distinguish a permutation from a vector.)
For example

$$[e_2, e_5, e_1, e_4, e_3]$$

is a permutation of the five elements $e_1, e_2, e_3, e_4, e_5$, in which

$$p_1 = e_2, p_2 = e_5, p_3 = e_1, p_4 = e_4, p_5 = e_3.$$

There is an objective function associated with any permutation $[P]$ which may be denoted by $F[P]$, and there may be constraints on the positions which the elements can occupy. These constraints will be embodied in a set of feasible permutations $G$. Then the problem is to determine a permutation $[P]$

$$[P] = [p_1, p_2, ..., p_N]$$

to minimize $F[P]$

subject to $[P] \in G$.

The following example illustrates a permutation formulation.

*Example 9.1*

Suppose $N$ jobs are to be scheduled through a single machine. Each job takes a certain amount of time on the machine, and has a due time by which it should be completed. Only one job can go on the machine at one time and a job cannot be interrupted once it is started. The due time of job $i$ is denoted by $t(i)$ and its duration by $d(i)$.

The cost of completing job $i$ say $x$ units late is $f(i, x)$. Let $p_j$ denote the $j$th job to be processed, then its completion time $c(p_j)$ is

$$c(p_j) = \sum_{k=1}^{j} d(p_k).$$

If the jobs are arranged in the order

$$[P] = [p_1, p_2, ..., p_N]$$

the total cost is

$$F[P] = \sum_{j=1}^{N} f(p_j, \max(c(p_j) - t(p_j), 0)).$$

The problem is to find the ordering which minimizes $F[P]$. There may also be constraints on the permutation in that certain jobs may have to precede others.

## 9.3 Locally optimal permutations

The problem of finding an optimal permutation is very difficult. In theory we could evaluate all possible solutions. But with $N$ elements, there are $N!$ possible solutions and this becomes a very large number. Even on a problem with 20 elements and a computer which could evaluate one solution every microsecond it would take more than 76,000 years to try all possibilities. The dynamic programming technique or the branch and bound method can sometimes be applied to small problems to reduce the search over the permutations, but often it is impossible to organize the problem in a stage-wise structure suitable for dynamic programming, and there may be no means of determining bounding formulae for the branch and bound method. Even where these methods can be applied their computational requirements may be excessive. We have reached the point at which it is necessary to trade off some of the quality of the answer for a reduction in the quantity of computation.

Instead of trying to find the globally optimal permutation we will aim simply to find a permutation which is better than some other 'near by' permutations. We have to define what we mean by 'near by'. Two permutations will be defined to be near by, or in the same neighbourhood, if one can be reached from the other with a few shuffles or interchanges of the ordering of the elements. For example the permutation $[P_1]$ of the first five integers

$$[P_1] = [4, 3, 1, 5, 2]$$

is comparatively 'near' the permutation $[P_2]$

$$[P_2] = [4, 5, 3, 1, 2]$$

as the permutation $[P_2]$ can be reached from the permutation $[P_1]$ simply by moving element number 5 to the second position in between elements 4 and 3 as shown in Fig. 9.1. On the other hand to reach the permutation

$$[1, 2, 3, 4, 5]$$

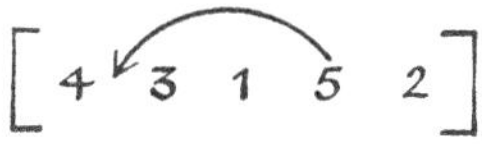

$$\left[ 4 \;\; 3 \;\; 1 \;\; 5 \;\; 2 \right]$$

Fig. 9.1

from $[P_1]$ we would have to perform three movements as indicated in Fig. 9.2

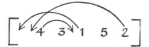

Fig. 9.2

moving 1 into the first position, 2 into the second position then 4 into the fourth position.

Rather than define a set of particular permutations to be the neighbourhood of each permutation $[P]$ we define the neighbourhood of a permutation $[P]$ in terms of a set of possible shuffles or exchanges on the elements in the permutation. If we denote the set of shuffles or exchanges by $S$ and arrange them in a list so that the $i$th exchange in the set is denoted by $E_i$, then the permutation $[P^{(i)}]$ is reached from $[P]$ by applying the exchange $E_i$ to the elements of $[P]$ and $[P^{(i)}]$ is obtained by the operation

$$[P^{(i)}] = [E_i P].$$

If there are $n$ exchanges in the set $S$ the neighbourhood of $[P]$ with respect to the set $S$ as defined by the permutation $[P^{(i)}]$, $i = 1, 2, ..., n$.

The simplest set of exchanges which we might consider would be the set of adjacent exchanges in which any one element and its adjacent element in the permutation are interchanged. Thus the $i$th exchange $E_i$ would denote the interchange of elements $p_i$ and $p_{i+1}$ and there would be $(N-1)$ exchanges in the list with $N$ elements. The four adjacent exchanges of a 5 element permutation denoted by $E_1, E_2, E_3, E_4$, could be applied to the permutation

$$[P] = [1, 2, 3, 4, 5]$$

to give the exchanged permutations as:

$$[E_1 P] = [2, 1, 3, 4, 5]$$

$$[E_2 P] = [1, 3, 2, 4, 5]$$

$$[E_3 P] = [1, 2, 4, 3, 5]$$

$$[E_4 P] = [1, 2, 3, 5, 4].$$

These four permutations define the neighbourhood of the permutation $[P]$ with respect to the set of exchanges $E_1, E_2, E_3, E_4$.

Clearly adjacent exchanges determine a very restricted neighbourhood of a permutation, and in practical problems we may choose a larger set of exchanges defining a more extensive neighbourhood. We will discuss the choice of the set of exchanges in a later section, but, generally, on the applications which we study in later chapters, we shall be concentrating on using exchanges in which we allow individual elements or consecutive blocks of elements to be moved to new positions in the permutations. A block shift exchange of this type is illustrated in Fig. 9.3 in which the consecutive set of elements $p_i, p_{i+1}, ..., p_{i+k}$ is moved to positions $j, j+1, ..., j+k$.

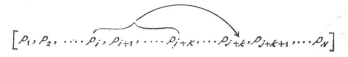

$$[p_1, p_2, .... p_i, p_{i+1}, .... p_{j+k} .... p_{i+k}, p_{j+k+1} .... p_N]$$

Fig. 9.3

In most of the applications which are considered later the size of the blocks are restricted to single elements so that the neighbourhood consists of all single shift exchanges.

The definition of the locally optimal permutation follows naturally from the definition of a neighbourhood. A permutation is locally minimal in the chosen neighbourhood if the objective function cannot be reduced by making any of the exchanges to the permutation. Using the same notation as before, if $S$ denotes the set of exchanges, the permutation $[\bar{P}]$ is locally optimal if

$F[\bar{P}] \leq F[E\bar{P}]$ for $E \in S$

where $[E\bar{P}] \in G$ (the set of feasible permutations).

The next example distinguishes a local optimum from a global optimum.

*Example 9.2*

We wish to find the shortest circuit through seven points in a plane which are located as shown in Fig. 9.4. The optimal circuit is clearly

1, 2, 3, 4, 5, 6, 7, 1.

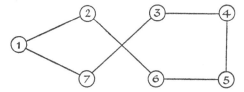

Fig. 9.4

Suppose we define the local optimum in terms of all single shift exchanges, where a single shift exchange is the movement of one element to a new position in the permutation. Then the permutation [1, 2, 6, 5, 4, 3, 7] (implying the circuit 1, 2, 6, 5, 4, 3, 7, 1) marked in the diagram is a locally optimal permutation because it is impossible to reduce the length of this circuit by moving any single element to a new position in the permutation, although it is clearly a longer circuit than the global minimum.

## 9.4  Calculation of locally optimal permutations

In common with other optimization methods, there are two stages to the task of obtaining a locally optimal permutation. The first stage constructs an initial feasible permutation, and the second stage modifies this initial permutation by applying a series of exchanges until an optimal permutation has been obtained. The overall strategy of the method is to obtain a locally optimal permutation with a minimum amount of computation.

The initial permutation may either be known in advance or may be constructed by a systematic method. A good rule for creating an initial permutation is to insert the elements into the permutation one by one. Once an element has been inserted into the permutation it remains a permanent part of it. If there are $N$ elements we need to make $N$ insertions to obtain a complete permutation. To apply this procedure, rules are needed for selecting the next element, choosing its initial position in the permutation, and adjusting the positions of the elements already in the permutation after each insertion. Different rules have to be devised to suit each application, but a good policy is to choose the next element as the one which most interacts with the partial permutation already constructed, and select its initial position so as to minimize the estimated increase in the objective function value. We shall be demonstrating this method for obtaining an initial permutation in later applications and in the next chapter we shall illustrate a heuristic scheme for calculating an initial tour on a travelling salesman problem. Because of the local nature of the optimum which is being obtained it is sometimes very important to try to obtain a good initial permutation from which the limited set of exchanges can work towards an answer reasonably close to the global optimum. Wherever possible any special properties or structure of the problem should be exploited to this end.

The second stage procedure for modifying the initial permutation to obtain a local optimum is very similar to the direct search procedure of non-linear programming. Starting with the initial permutation we apply the exchanges in turn and whenever an exchange could lead to a feasible permutation and to a reduction in the objective function value we accept the exchange. Otherwise we leave the permutation unchanged. In either event we continue the procedure with the next exchange in the list, until we reach the last exchange whereupon we return to the first exchange. We refer to the best permutation obtained so far as the base permutation and we will call the trial permutations, which are reached from this base, tentative permutations. As exchanges are only being made when they reduce the objective function value the objective function values associated with the series of bases will necessarily get steadily smaller, and the procedure will ultimately converge to a local minimum. This occurs when a full cycle of exchanges has been made with no objective function value reduction.

Suppose the set $S$ of exchanges defining the local optimum contains $n$ exchanges $E_1$, $E_2$, ..., $E_n$. Let $[P^{(0)}]$ denote the initial permutation which is also the first base permutation and let $[P^{(t)}]$ with elements

$$[P^{(t)}] = [p_1^{(t)}, p_2^{(t)}, ..., p_N^{(t)}]$$

be the permutation being tried at the $t$th iteration. Also we will denote by $[B^{(t)}]$ the base permutation at iteration $t$ which is the best permutation determined up to this point. At the first iteration where $[B^{(0)}] = [P^{(0)}]$ we calculate

$$[P^{(1)}] = [E_1 B^{(0)}].$$

If

$F[P^{(1)}] < F[B^{(0)}]$ and $[P^{(1)}] \in G$
$[B^{(1)}] = [P^{(1)}]$

otherwise

$[B^{(1)}] = [B^{(0)}]$.

Next we calculate $[E_2 B^{(1)}]$ and if this leads to a feasible permutation which reduces the objective function it is accepted as the new base. Otherwise the base remains $[B^{(1)}]$.

In general if, at iteration $t$, exchange $E_{t-1}$ was applied at the previous iteration we calculate

$[P^{(t)}] = [E_t B^{(t-1)}]$

and if $F[P^{(t)}] < F[B^{(t-1)}]$ and $[P^{(t)}] \in G$

$[B^{(t)}] = [P^{(t)}]$

otherwise

$[B^{(t)}] = [B^{(t-1)}]$.

(It should be noted that $E_k = E_1$ if $E_{k-1} = E_n$.)

The permutation $[B^{(T)}]$ at iteration $T$ is a local minimum with respect to the set of exchanges $S$ if

$F[B^{(T)}] = F[B^{(T-n)}]$.

The exchanging procedure which has been proposed is not the only method which could be used to obtain the locally optimal permutation. Instead of cycling serially through the exchanges without regard to the occurrence of improvements we could return to the first exchange each time an improvement occurred. However, it is usually worth while to operate on the whole permutation with equal intensity and we therefore proceed through the exchanges cyclically. Another scheme would be to examine what would be the effect of trying all possible exchanges on any given permutation and choose the one which would lead to the greatest reduction. But in practice it has been found that there is no real difference in the quality of the locally optimal permutation obtained and the computation time required is very much larger.

*Example* 9.3

In a two-machine scheduling problem similar to the three-machine scheduling problem considered in Section 8.3, all jobs have to proceed through each of two machines $A$ and $B$ in the same order $A$, $B$. The durations of the four jobs are shown in the Table 9.1. The objective is to sequence the jobs through the machines so as to minimize total production time. We denote a permutation of the four jobs by

$[P] = [p_1, p_2, p_3, p_4]$.

If $x_a(p_j)$, $x_b(p_j)$ denote the start times of the $j$th job on machines $A$ and $B$ respectively

$F[P] = x_b(p_4) + b(p_4)$

F

where $x_a(p_j)$ and $x_b(p_j)$ are calculated recursively as

$x_a(p_j) = x_a(p_{j-1}) + a(p_{j-1})$ and $x_a(p_0) = 0$

$x_b(p_j) = \max (x_a(p_j) + a(p_j), x_b(p_{j-1}) + b(p_{j-1}))$, and $x_b(p_0) = 0$.

The formulae are similar to those used for Section 8.3 in the branch and bound method.

We will choose all adjacent exchanges in which two adjacent elements in the permutation may be interchanged as the set for the local optimum. For the initial permutation, we will arrange the jobs in order of increasing total processing time on the two machines. This gives an initial permutation as [1, 2, 3, 4] and a total production time of 19.

**Table 9.1**

| Job $j$ | Time on $A$ $a(j)$ | Time on $B$ $b(j)$ |
|---------|--------------------|--------------------|
| 1 | 2 | 3 |
| 2 | 4 | 2 |
| 3 | 5 | 2 |
| 4 | 2 | 6 |

We now perform the second stage procedure on this initial permutation. The three adjacent exchanges will be listed as

$E_1$ the interchange of jobs $p_1$ and $p_2$,

$E_2$ the interchange of jobs $p_2$ and $p_3$,

$E_3$ the interchange of jobs $p_3$ and $p_4$.

The sequence of permutations, their costs and the changes of base are then determined as follows:

| Exchange | Permutation | Production time | Base |
|----------|-------------|-----------------|------|
| | $[P^{(0)}] = [1, 2, 3, 4]$ | 19 | $P^{(0)}$ |
| $E_1$ | $[P^{(1)}] = [E_1 P^{(0)}] = [2, 1, 3, 4]$ | 19 | $P^{(0)}$ |
| $E_2$ | $[P^{(2)}] = [E_2 P^{(0)}] = [1, 3, 2, 4]$ | 19 | $P^{(0)}$ |
| $E_3$ | $[P^{(3)}] = [E_3 P^{(0)}] = [1, 2, 4, 3]$ | 16 | $P^{(3)}$ |
| $E_1$ | $[P^{(4)}] = [E_1 P^{(3)}] = [2, 1, 4, 3]$ | 17 | $P^{(3)}$ |
| $E_2$ | $[P^{(5)}] = [E_2 P^{(3)}] = [1, 4, 2, 3]$ | 15 | $P^{(5)}$ |
| $E_3$ | $[P^{(6)}] = [E_3 P^{(5)}] = [1, 4, 3, 2]$ | 15 | $P^{(5)}$ |
| $E_1$ | $[P^{(7)}] = [E_1 P^{(5)}] = [4, 1, 2, 3]$ | 15 | $P^{(5)}$ |
| $E_2$ | $[P^{(8)}] = [E_2 P^{(5)}] = [1, 2, 4, 3]$ | 16 | $P^{(5)}$ |

Thus the permutation [1, 4, 2, 3] is locally optimal with respect to adjacent exchanges and the total production time is 15.

## 9.5 Choosing sets of exchanges

The most difficult question in using a permutation procedure is to choose a suitable set of exchanges to define the local optimum. This is where the application of the method raises most uncertainties. Ideally we want to find a small set of exchanges which, applied through the second stage procedure, will determine the global optimum starting from any initial permutation. Gomory and Maxwell have found some examples in scheduling where this can be done, but these are rare cases, and we are commonly faced with the issue of choosing the set of exchanges which will solve the problem satisfactorily. What is satisfactory is a matter of judgement, but clearly we require local optima which are close to the global optimum. Measuring this proximity may be very difficult and will often involve experimental and statistical investigations. For example, we can start from a number of random initial permutations, use the second stage procedure to obtain a sequence of local optima and stop sampling when the cost of further trials outweighs the potential return. Another possibility is to steadily expand the set of exchanges starting from one initial permutation until the result is not being improved upon significantly. A further scheme is to compare the quality of answers obtained by a variety of exchanges on artificially small problems where the optima are known by other methods such as branch and bound or dynamic programming.

In practical terms these difficulties may be less severe than they appear to be from an academic viewpoint. It is often possible to interpret the physical meaning of exchanges on permutation problems and then the problem solver may decide in advance what constitutes a satisfactory local optimum. For example single shift exchanges may be appropriate for a scheduling problem if the production planner requires a schedule which cannot be improved by the manipulation of any individual job to a new position in the schedule. With several hundred jobs in the schedule this will be much more than a manual planner could achieve and hence should offer substantial gains. Or in circuit design, the placement of components so that no improvement can be obtained from a pair interchange may be a good deal better than the layouts which a visual design procedure will achieve. Often an interpretation of the problem in this way will automatically designate an appropriate set of exchanges, and this will be demonstrated on later applications.

## *9.6 Multi-permutation problems

Although the permutation representation provides an adequate framework for many sequencing problems, there are more complex situations which need to be represented in terms of a number of permutations. In these cases the elements have to be allocated to the permutations as well as sequenced within the permutation. If the $i$th permutation is denoted by $[P_i]$ and it contains $n(i)$ elements it could be denoted by

$$[P_i] = [p_{i1}, p_{i2}, \ldots, p_{in(i)}].$$

For example the two circuits illustrated in Fig. 9.5 would be represented by the two permutations

$$[P_1] = [1, 2, 3, 6, 7, 10, 11]$$

and

$$[P_2] = [4, 5, 9, 8].$$

An objective function might be to find the two circuits whose combined lengths are minimized. The objective function would be expressed in terms of

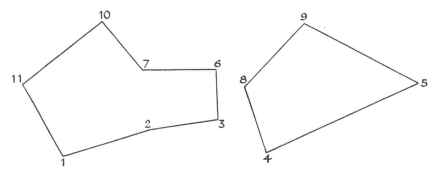

Fig. 9.5

both permutations as $F[P_1, P_2]$. The same kind of optimization procedure could be applied to these problems and exchanges of the elements between the permutations could be considered as well as within the permutation. We shall be referring to some multi-permutation problems in the applications.

REFERENCES

Gilmore, P. C. and Gomory, R. E. 1964. Sequencing a one state-variable machine: a solvable case of the travelling salesman problem. *Opns. Res.*, **12**, 655.

Maxwell, W. L. 1964. The scheduling of single machine systems: a review. *Int. J. Prod. Res.*, 3, 177.

Nicholson, T. A. J. 1967. A method for discrete optimization problems and its application to the assignment travelling salesman and three machine scheduling problems. *J. Inst. Math. Appl.*, 3, 362.

Reiter, S. and Sherman, G. 1965. Discrete optimizing. *J. Soc. Ind. Appl. Math.*, **13**, 864.

## Exercises on Chapter 9

**1** Six points in a plane lie at coordinates $(-1, 0)$, $(0, 0)$, $(1, 0)$, $(-1, 1)$ $(0, 1)$, $(1, 1)$. Find a circuit through these points which is locally minimal with respect to adjacent exchanges but not globally minimal.

**2** Apply the permutation procedure of Example 9.3 to the two machine scheduling problem of Exercise 8.1.

**3** In a transport system 5 trucks are to be passed through a single loading bay and there is a set-up time if truck $j$ follows truck $i$ as given in Table 9.2.

**Table 9.2**

|  |  | | | $i$ | | |
|---|---|---|---|---|---|---|
|  |  | 1 | 2 | 3 | 4 | 5 |
|  | 1 | 0 | 4 | 2 | 6 | 5 |
|  | 2 | 6 | 0 | 7 | 2 | 2 |
| $j$ | 3 | 0 | 4 | 0 | 4 | 6 |
|  | 4 | 2 | 3 | 2 | 0 | 3 |
|  | 5 | 7 | 3 | 1 | 3 | 0 |

(The first truck at the loading bay can start at time 0.) The basic loading time for all the trucks is 4 units of time and the objective is to minimize the total time required to load all the trucks at the loading bay. Starting with the permutation [1, 2, 3, 4, 5] perform adjacent exchanges to obtain a local optimum with respect to all adjacent exchanges.

**\*4.** Suppose that we have exactly the same problem structure as was considered for Example 9.1 except that there are $N$ jobs each with one operation to go through any of 3 machines. Assume also that the durations of the jobs depend on which machines the job proceeds through. Formulate this problem as a multi-permutation problem.

**\*5** $N$ connected elements numbered 1 to $N$ are to be placed anywhere on a fixed circle. The connections between the elements are given by a connection matrix in which

$C(i, j) = 1$ if elements $i$ and $j$ are connected

$\quad\quad\quad = 0$ otherwise.

Assuming that the connection between two points is drawn as a straight line formulate the problem of minimizing the number of crossings between the connections as a permutation problem.

# 10 Heuristic techniques

## 10.1 The need for skilful guessing

In this final chapter on optimization procedures we will briefly examine the nature of a class of methods to which operational research nomenclature has assigned the obscure name of 'heuristic techniques'. These are methods for solving problems by an intuitive approach in which the structure of the problem can be interpreted and exploited intelligently to obtain a reasonable solution. Heuristic techniques are the only way of tackling many large optimization problems. Examples occur in railway timetabling, resource allocation in large interdependent projects, or the location of plants and warehouses in complex overlapping markets. Although we may have a criterion for these problems and be aware of what constitute the problem variables and the restrictions we still cannot express them in suitable terms to apply a formal optimization technique. But the problems have to be solved: trains are time-tabled, resources are allocated on building sites, and plants are located near market centres without the help of formal methods. They are tackled by skilful guesswork in which a series of intelligent decisions are made to determine suitable values for the problem variables and the answers are accepted as adequate. The intelligence which goes into the decision making often embodies the accumulated experience of individuals who have solved the problem many times, and their intuitive understanding can be of real benefit in the solution.

The generality and wide application of heuristic techniques makes it impossible to present a single framework for these methods, but in the following sections we shall outline a heuristic approach to a class of allocation problems where heuristic ideas have been widely applied. Two points should be stressed about the nature of the method. First we will view a heuristic technique as a means of making a series of intelligently chosen decisions to devise a solution. (The intelligence is thus embodied in the way the problem variables are assigned values rather than about the type of answer required as was the case with the local optimum of a permutation problem.) Secondly, because the decisions are only made once, it is essentially a first-stage procedure for obtaining an initial solution without any second-stage improvement. Consequently a heuristic method is often computationally quick although it will give no measure of the quality of the answer.

## 10.2  The nature of allocation problems

Heuristic techniques have been used extensively on awkward allocation problems. Let us characterize an allocation problem as the assignment of a set of values to a number of items. The values may be the times at which trains start from their departure points or the location of production centres on a map. If there are $N$ items, the values assigned to the items will be denoted by the $N$ problem variables $x_1, x_2, ..., x_N$. There is an overall objective function $F(x_1, x_2, ..., x_N)$ associated with any assignment and there may be very complex constraints interrelating the assignments. Furthermore, it is assumed that it is possible to attach an estimated objective function value to a partial solution in which some of the assignments have been made but not all, and that the constraints can be contemplated naturally in terms of partial solutions. This makes it possible to build up the solution in a sequential fashion assigning a value to one variable at a time. The basis of the heuristic techniques which we will describe is to exploit this feature and to solve these problems by making a series of $N$ decisions—each decision selecting a variable and assigning a value to it. It should be stressed that this is only one way in which heuristic ideas are employed.

Suppose that after $t$ decisions have been made values have been assigned to the $t$ variables $x_{s(1)}, x_{s(2)}, ..., x_{s(t)}$ and that their values are denoted by $\bar{x}_{s(1)}, \bar{x}_{s(2)}, ..., \bar{x}_{s(t)}$. We now select one of the variables which has not yet been assigned a value and choose a value for it. This is a two-stage decision. First, we choose the next item to be assigned. For this purpose an item preference or cost measure is used for each unassigned item $i$ as

$$g(s(1), s(2), ..., s(t); i).$$

The next item $i$ is chosen to maximize or minimize this measure. Usually the measure will be denoted simply by

$$g_t(i).$$

Secondly, we choose a value for the selected item by minimizing (or maximizing) the value measure for item $i$

$$f(\bar{x}_{s(1)}, \bar{x}_{s(2)}, ..., \bar{x}_{s(t)}; x_i).$$

This generally measures the cost of assigning the value $x_i$ to the $i$th item after the assignment

$x_{s(k)} = \bar{x}_{s(k)}$ for $k = 1, ..., t$

has been made.

Again this measure may be denoted simply by

$f_t(x_i)$.

After $N$ iterations of the procedure making one assignment each time, a complete solution

$(\bar{x}_1, \bar{x}_2, ..., \bar{x}_N)$

has been determined and this is accepted as the answer.

Although this heuristic technique for solving allocation problems will generally solve the problems quickly—which is its chief virtue—it will give no idea how close to the optimum the final answer lies. Sometimes the solution will be a long way from the optimum. The decisions are made irrevocably, each successive decision being based on an approximate assessment of future conditions and being committed by the previous decisions. There is a tendency therefore for the assignments to become more difficult and less efficient as the decisions sequence proceeds. In order to minimize these effects it is important to exploit as fully as possible any special structure in a problem.

Three examples of the use and need for heuristic procedures will now be presented. The first problem is very elementary and can be solved optimally by other methods, but its solution by a heuristic procedure helps to illustrate the basic ideas. The second example is the familiar travelling salesman problem. The third example on timetabling is much more difficult, and we will not give a detailed solution; it demonstrates the need to tackle a complex problem by a heuristic decision-making process.

*Example* 10.1

Five items numbered 1 to 5 are to be assigned to 5 positions, also numbered 1 to 5, and the cost of assigning item $i$ to position $j$ is denoted by $c(i, j)$ as given in Table 10.1.

**Table 10.1**

|  |  | Position $j$ |  |  |  |  |
|---|---|---|---|---|---|---|
|  |  | 1 | 2 | 3 | 4 | 5 |
|  | 1 | 18 | 11 | 9 | 10 | 5 |
|  | 2 | 12 | 14 | 6 | 13 | 19 |
| Item $i$ | 3 | 4 | 5 | 4 | 3 | 2 |
|  | 4 | 17 | 15 | 12 | 18 | 9 |
|  | 5 | 19 | 10 | 14 | 11 | 6 |

The problem is to allocate an item to each position so as to minimize the total cost. Only one item can be assigned to each position. This is the well-known assignment problem and it can be solved by linear programming methods. However, it is illustrative to tackle it with a heuristic technique.

A solution to the problem is denoted by $(x_1, x_2, x_3, x_4, x_5)$ where $x_j$ is the position to which item $j$ is assigned. A heuristic technique for solving this problem is to choose the next item as the one which has the least remaining available cost and allocate it to the position for which this cost is a minimum. Writing $g_t(i)$ as the least cost of assigning item $i$ after $t$ items have been assigned, we get

$$g_0(1) = \min (18, 11, 9, 10, 5) = 5$$
$$g_0(2) = \min (12, 14, 6, 13, 19) = 6$$

and similarly,

$$g_0(3) = 2, g_0(4) = 9, g_0(5) = 6.$$

**Table 10.2**

|   | 1 | 2 | 3 | 4 |
|---|---|---|---|---|
| 1 | 18 | 11 | 9 | 10 |
| 2 | 12 | 14 | 6 | 13 |
| 4 | 17 | 15 | 12 | 18 |
| 5 | 19 | 10 | 14 | 11 |

Item 3 is selected as it has the least $g_0(i)$ measure. The costs for its various positions are:

$$f_1(x_3 = 1) = 4, f_1(x_3 = 2) = 5, f_1(x_3 = 3) = 4, f_1(x_3 = 4) = 3,$$
$$f_1(x_3 = 5) = 2.$$

Item 3 is therefore allocated to position 5 for which $f_1(x_3)$ is a minimum. This contributes a cost of 2 units.

We now have a 4 by 4 cost matrix as shown, in Table 10.2. For this iteration

$$g_1(1) = 9, g_1(2) = 6, g_1(4) = 12, g_1(5) = 10.$$

Hence item 2 is selected and assigned to position 3 with cost 6. At the third iteration

$$g_2(1) = 10, g_2(4) = 15, g_2(5) = 10.$$

Hence item 1 is selected and assigned to position 4 with cost 10. At the fourth iteration

$$g_3(4) = 15, g(_35) = 10.$$

Hence item 5 is assigned to position 2 with cost of 10. Finally, item 4 is allocated to position 1 with cost 17.

The total cost of this assignment is $2+6+10+10+17$, giving a total of 45. The minimum cost, on the other hand, is 39 and is achieved by allocating the items to positions 4, 3, 1, 5, 2 respectively. The results illustrate how a heuristic procedure can lead to poor decisions towards the end of the sequence.

This is only one heuristic rule for the assignment problem. Other schemes are equally possible. For instance, instead of selecting the item with the least cost position which remains we could select the item with the largest least cost position which remains and assign it to its cheapest position. This rule might be justified on the grounds that it is more important to get the necessarily expensive items allocated first. When this scheme is used it leads to the sequence of allocations:

item 4 allocated to position 5 with cost 9

item 5 allocated to position 2 with cost 10

item 1 allocated to position 3 with cost 9

item 2 allocated to position 1 with cost 12

item 3 allocated to position 4 with cost 3,

giving a total cost of 43 units which is 2 units better than the first scheme.

A third heuristic for the assignment problem is examined in Exercise 10.1. It is important to realize the scope for alternative heuristics. The schemes will not necessarily be consistent: scheme 1 may be better than scheme 2 on one occasion and worse on another. But generally it is possible to distinguish good and bad heuristics by making a number of experimental trials.

*Example* 10.2

The travelling salesman problem provides an opportunity for indulging in heuristic techniques. Suppose we have $N$ points in a plane numbered 1, 2, ..., $N$ with distance $d(i, j)$ from point $i$ to point $j$. We wish to find the ordering of the points say $(i_1, i_2, ..., i_N)$ which minimizes the total distance around the tour. We can view the problem in terms of choosing $N$ cities for $N$ positions in the tour. For the first decision we choose say city 1, and place it in position 1. Next we choose a city for position 2 and we can choose the city as the one which is the least distance away from city 1. We can continue in this way choosing the next city as the one which is the least distance from the most recent city to be inserted. This heuristic is called the 'nearest city' method. After $t$ cities have been chosen to form the first $t$ positions in the tour as

$(i_1, i_2, ..., i_t)$

the next city to be chosen is the one which has so far not been selected and is the least distance from city $i_t$. Thus the city $i_{t+1}$ is chosen as the one which minimizes the measure

$$g_{t+1}(i) = d(i_t, i)$$

where $i \neq i_1, i_2, ..., i_t$. The city $i_{t+1}$ which is chosen is placed in position $(t+1)$.

To illustrate this approach, suppose the data of Table 10.3 represents the distances between 8 cities. The element $(i, j)$ is the distance between city $i$

**Table 10.3**

|     |     |     |     | $j$ |     |     |     |     |
|-----|-----|-----|-----|-----|-----|-----|-----|-----|
|     | 1   | 2   | 3   | 4   | 5   | 6   | 7   | 8   |
| 1   | 0   | 4   | 7   | 11  | 2   | 6   | 3   | 8   |
| 2   | 4   | 0   | 1   | 5   | 4   | 8   | 4   | 4   |
| 3   | 7   | 1   | 0   | 17  | 9   | 16  | 13  | 10  |
| $i$ 4 | 11  | 5   | 17  | 0   | 8   | 12  | 7   | 6   |
| 5   | 2   | 4   | 9   | 8   | 0   | 5   | 11  | 6   |
| 6   | 6   | 8   | 16  | 12  | 5   | 0   | 16  | 14  |
| 7   | 3   | 4   | 13  | 7   | 11  | 16  | 0   | 8   |
| 8   | 8   | 4   | 10  | 6   | 6   | 14  | 8   | 0   |

and city $j$. Then if we start from city 1 we will go on to city 5 as this is the least distance from 1 giving the incomplete tour

(1, 5) with distance 2.

Next we go on to city 2 as this is the least distance from 5 amongst the excluded cities giving:

(1, 5, 2) with distance 6.

Next city 3 is selected:

(1, 5, 2, 3) with distance 7.

Next city 8 is selected:

(1, 5, 2, 3, 8) with distance 17.

Next city 4 is selected:

(1, 5, 2, 3, 8, 4) with distance 23.

Next city 7 is selected:

(1, 5, 2, 3, 8, 4, 7) with distance 30.

Finally, city 6 is selected as the only one left adding a distance 16 on to the tour with an additional 6 units to return from 6 to 1. The total distance is 52 for the tour

(1, 5, 2, 3, 8, 4, 7, 6, 1).

The difficulty with the nearest city approach is that the last cities to be added to the tour are likely to increase it out of proportion to the remainder as they are committed to go into positions in the tour which may not be suitable. Also no consideration is being given to the circular nature of the tour. A better procedure is to build up sub-tours by introducing one city at a time into the tour at its best position and considering the circular tour distance at each stage. The precise rule of selection and positioning is then to choose the next city as the one which, when it is placed in its best position in the current sub-tour minimizes the increase in tour length. If the current sub-tour is

$$(i_1, i_2, ..., i_t)$$

we choose the city $i$, not yet included, and place it in position $j$ if the circular tour length

$$d(i_1, i_2) + d(i_2, i_3) + ... + d(i_{j-1}, i) + d(i, i_j) + ... + d(i_{t-1}, i_t) + d(i_t, i_1)$$

for the tour

$$(i_1, i_2, ..., i_{j-1}, i, i_j, ..., i_t)$$

is a minimum for all possible $i$ and $j$. This implies that the cities in positions $i_j, ..., i_t$ are being shuffled along one place in the sequence if the next city is placed in position $j$. This method will in general yield much better results than the nearest city scheme. A further good heuristic which is suitable for planar travelling salesman problems is considered in Exercise 10.4.

*Example* 10.3

Curriculum timetabling in schools and universities is a complex allocation problem for which no general satisfactory method yet exists. Many of the techniques which have so far been proposed for solving these problems are based on heuristic schemes.

The timetabling problem is to assign classes to the periods of a weekly timetable (and also allocate classrooms and teachers) so as to meet some desired objective. The available periods of the week may be listed as 1, 2, ..., $T$ which for simplicity may be considered to be all of equal duration. A class consists of a combination of one or more sets of students, a suitable classroom and a teacher. (Often the teacher is assigned to the class in advance.) The data for the problem, therefore, consists of a list of class requirements, a list of sets of students, a list of available rooms and their capacities, and a list of teachers. The classes are to be allocated time periods in the range 1, ..., $T$, where class $i$ requires a set $S_i$ of students, a room type $R_i$ and a teacher type $Y_i$. Each class requirement is to be assigned a period $x_i$ in the course of the week.

The objective usually includes a mixture of factors. The predominant objective is that all the classes should be able to be included in the timetable,

A secondary objective is to ensure a reasonable distribution of the subjects over the working week.

The constraints are that no set of students, teacher, or room is allocated twice at the same period. (There may in practice be additional constraints that certain classes have to occupy adjacent periods as a double, or even triple, period.)

This is a very complex allocation problem. Nevertheless, timetables are worked out frequently by experienced school teachers with great efficiency. It would be desirable to try to build their procedures into efficient heuristic schemes which could calculate timetables automatically. The stress in the schemes so far proposed is to maximize the number of classes which are inserted into the timetable, regardless of distribution, and we will consider a heuristic method for doing this.

At each decision a class is selected and a time period assigned to it. The choice of the class corresponds to the selection of the item for assignment and the time period is the value chosen for the item. The rules for selecting the item and assigning a time period generally aim to minimize the difficulty of assigning time periods to future unassigned classes.

The next class may be chosen as the class for which there is a minimum number of remaining feasible periods—a feasible period being one in which the sets of students, a suitable room and a teacher are all available when some assignments have already been made. If five classes had 5, 4, 10, 14, 8 possible periods left, we would choose the class with only 4 possible periods.

The next task is to decide which period should be chosen for the class. A sensible policy would be to allocate the class to the time period for which fewest of the other unassigned classes could be allocated. If there were four possible times and the number of unassigned classes which could be placed in each of these periods were 7, 3, 24, 11, we would allocate it to the period with only three competitors.

Although this policy is sensible it has still not successfully solved many timetabling problems in practice. Each decision allocates a set of classes, a teacher and a room to a time period; this restricts future decisions, and it has been found very difficult to ensure that all classes are inserted into the time-table. The final classes do not get a place. Sometimes it may be impossible to include all classes in the weekly period and no heuristic scheme would overcome this. It may be that practical timetablers adapt their requirements as they proceed with the allocation. It has been shrewdly suggested that the difficulties in timetabling problems only became apparent when attempts were made to analyse and solve them by computer using optimization procedures.

REFERENCES

Cooper, L. 1964. Heuristic methods for location-allocation problems. *Soc. Ind. Appl. Math. Rev.*, **6** (1), 37.

Feldman, E., Lehrer, F. A. and Ray, T. L. 1960. Warehouse location under continuous economies of scale. *Man. Sci.*, **12** (6).

Folkers, J. S. 1968. Time-table construction by computer. Paper at European Meeting, The Institute of Management Science, Amsterdam (1968).

Gere, W. S. 1966. Heuristics in job-shop scheduling. *Man. Sci.* **13** (3), 167.

Nicholson, T. A. J. 1968. A boundary method for the travelling salesman problem. *Op. Res. Q.*, **19**.

Tietz, M. B. and Bart, P. 1968. Heuristic methods for estimating the generalized vertex median of a weighted graph. *Opns. Res.*, **16** (5), 955.

## Exercises on Chapter 10

**1** Solve the assignment problem of Example 10.1 by the following scheme. After $t$ items have been assigned to positions assess the difference between the cheapest and second cheapest remaining costs for the unallocated items. Choose the item with the largest difference and assign it to its cheapest position. The case for this heuristic is that it stresses the costs of losing a reasonable position for an item for a very poor position, rather than concentrating on getting good positions for items which may have good alternative positions.

**2** A well-known transportation problem is to allocate units from supply points to demand points in the cheapest way. Suppose there are 4 supply points with supplies 5, 6, 2 and 9 respectively and 6 demand points with demands 4, 4, 6, 2, 4, 2 respectively. The costs of sending 1 unit from any supply point to any demand point is given in Table 10.4. Suggest a heuristic scheme for allocating the supplies to the demands and apply it to the given data.

**Table 10.4**

|  |  |  | Demand points | | | | | |  |
|---|---|---|---|---|---|---|---|---|---|
|  |  | Supplies | 1<br>4 | 2<br>4 | 3<br>6 | 4<br>2 | 5<br>4 | 6<br>2 | Demands |
|  | 1 | 5 | 9 | 12 | 9 | 6 | 9 | 10 |  |
|  | 2 | 6 | 7 | 3 | 7 | 7 | 5 | 5 |  |
| Supply points | 3 | 2 | 6 | 5 | 9 | 11 | 3 | 11 |  |
|  | 4 | 9 | 6 | 8 | 11 | 2 | 2 | 10 |  |

**3** Six points in a plane lie at coordinates (0, 0) (1, 1) (2, 0) (3, 1) (4, 0) and (5, 1). We wish to find a circuit through these points so as to minimize its length. Apply the nearest city method discussed in Example 10.2 to this data and also the improved method suggested at the end of Example 10.2.

**4** A special feature of planar travelling salesman problems (where all points lie in a plane) is that the sequence of points forming the convex boundary will be a subsequence of the minimal circuit. Confirm this result and suggest how it might be used to obtain a good complete tour.

# 11 Problem specification and mathematical treatment

## 11.1 Flexibility in problem formulation

The success of any optimization study depends largely on the quality of the problem formulation. This is undoubtedly the most important and most difficult task in applying the techniques. There are two stages to a formulation. First the application has to be identified, isolated and modelled. Secondly, after deciding on the general shape of the problem it has to be expressed as a mathematical opimization problem. In both tasks there are a number of important decisions to make and considerable room for manœuvre. In this chapter we will review the flexibilities which can be exploited in the mathematical definition, and in Chapter 12, beginning Volume 2, we examine the issues in the construction of models.

In all the examples of optimization problems which we have studied so far there has never been any real difficulty about identifying what the problem variables were and how the objective and constraint functions were composed. In practice, life is not as straightforward as this, usually there are a number of options. There may be alternative ways of choosing the problem variables, and the factors which go into constraints and into the objective function may to some extent be interchangeable. It may be advantageous to introduce auxiliary variables to re-express the constraints in a different form. Alternatively we may be able to transform the problem mathematically to eliminate the constraints. Depending on how we make these decisions, we will end up with different forms of optimization problem and consequently be able to use different techniques. A special flexibility arises in linear programming where we may advantageously investigate a 'dual' problem which although it appears to be a completely different treatment will produce the same answers and yield additional information as well. We will review some of these possibilities in the following sections.

The case for recognizing and exploiting this mathematical flexibility arises from the desirability of ending up with a problem which can be solved by a standard method. It should be recalled how some of the optimization techniques require a problem to be viewed from a distinctive standpoint. Dynamic programming requires a special multi-stage interpretation of a problem, and the branch and bound procedure examines the solution in terms of a tree search. If the functions are nearly linear we may be able to approximate the functions in order to use linear programming, or if they are well-behaved non-linear functions it may be possible to adjust them to suit the methods of separable or approximation programming. All these possibilities must be considered when an optimization problem is defined mathematically.

## 11.2 The choice of the problem variables

Once the physical problem has been formulated there is generally little doubt about what constitutes the problem variables. They may be production levels, the numbers of items to hold in stock, the types of plant which should be used in a replacement plan or the settings on the valves of a chemical process. They are the quantities which we are able to control. However, there may be a choice about how they are defined mathematically. We may have a discrete variable in the original problem which can be treated as a continuous variable if the rounding error can be accepted. This significantly affects the optimization techniques which are available for the solution. Equally, in sequencing operations through a machine, the sequence can be specified either by the set of start times on the machine or as a permutation of the operations. In some cases it may be right to represent the start times as continuous variables and perhaps use linear programming, but in other contexts a sequence representation and a permutation procedure would be more appropriate. In a stock control scheme it may be better to treat the aggregate stock taken on up to various times as the basic variables, rather than the individual stock intakes. Fortunately the selection of the problem variables is not a field in which we are likely to go too far astray. Provided the alternatives are recognized there is usually a best form for the problem variables which manifests itself automatically.

## 11.3 Transferring factors between the objective and constraint functions

The distinction between objectives and constraints is not always a precise one, and we may be able to exploit this haziness by transferring constraints to the objective function or expressing part of the objective as a constraint. Generally speaking, a requirement is a constraint if it must not be violated at any cost and there is no gain from over-fulfilling it, and a requirement is an objective if it can be violated at a cost and there is an advantage in over-fulfilling it. For instance the constraint

$$x \leqq 5$$

is unaffected whether $x$ is 0 or 4 but the solution $x = 6$ is ruled out. Whereas if we were maximizing $x$ it is better to obtain $x = 4$ than $x = 0$ and still better to reach $x = 6$. But seldom are the requirements clearly divided into these categories of objectives and constraints.

The idea of linking up the constraints and objective function has been met already in the created response surface technique for non-linear optimization where the constrained problem is converted into an unconstrained form by building the constraints into the objective function so that they become penal-ties if they are violated. But there are alternative ways of doing this and the schemes should be interpreted directly in terms of the problems as one may be more appropriate than another. The following two examples illustrate the transfer ideas.

*Example* 11.1

The design of a product is specified in terms of three variables $x_1$, $x_2$, $x_3$. The product has two properties which depend on $x_1$, $x_2$, $x_3$ and whose values are specified as $g(x_1, x_2, x_3)$ and $h(x_1, x_2, x_3)$. The values of the properties are constrained to be greater than $a$ and $b$ respectively, so that we require

$$g(x_1, x_2, x_3) \geqq a$$

$$h(x_1, x_2, x_3) \geqq b.$$

It is required to find the $x_i$ values which minimize a cost function $F(x_1, x_2, x_3)$ subject to satisfying the constraints. The response surface idea of Chapter 6 suggests that the constraints should be incorporated into the objective function to create a new objective function

$$P(x_1, x_2, x_3, r) = F(x_1, x_2, x_3) + \frac{r}{g(x_1, x_2, x_3) - a} + \frac{r}{h(x_1, x_2, x_3) - b}$$

where $r$ is some suitably chosen parameter.

Another method for incorporating the constraints into the objective func-tion is to add them on as linearly increasing penalties the more they are violated. This creates the function $Q(x_1, x_2, x_3)$:

$$Q(x_1, x_2, x_3, r) = F(x_1, x_2, x_3) + r \max (a - g(x_1, x_2, x_3), 0)$$

$$+ r \max (b - h(x_1, x_2, x_3), 0)$$

where $r$ is a large number and the constraints become active components of the objective function only when they are violated. The direct search method could be used to minimize the function $Q$.

Although it is often mathematically desirable to make this kind of transfer it can lead into difficulties of deciding the weights for combining different quantities normally measured in different units. Once a constraint is incorporated into a cost objective function it is measured as a cost and we are effectively allowing it to be violated—at a penalty. Therefore we need to choose weights ($r$ values) so as to ensure that it is never worthwhile to violate a constraint for the sake of further reductions in the original cost measure. If a technical restriction is embodied in an objective function otherwise composed of economic factors it may be quite difficult to judge the correct weights.

There are also contexts in which a component of the objective function may be transferred to a constraint. This can have a simplifying effect when defining the problem if some of the components of the objective function are not well specified. Example 11.2 provides an illustration.

*Example* 11.2

Production levels are to be determined over $N$ periods so as to minimize manufacturing costs and maximize operator utilization. Let $x_i$ denote the production level for the $i$th period, let $f_i(x_i)$ denote the manufacturing cost in period $i$ and let $g_i(x_i)$ be the operator utilization returns at the level $x_i$. The objective function may be expressed as

$$F(x_1, x_2, ..., x_N) = \sum_{i=1}^{N} (f_i(x_i) + g_i(x_i)).$$

However, the functions $g_i(x_i)$ may be difficult to define and instead we may be able to constrain operator utilization to be within a certain range, say $(l(i), L(i))$ in period $i$. The new problem becomes

$$\text{minimize} \sum_{i=1}^{N} f_i(x_i)$$

subject to $l(i) \leq g_i(x_i) \leq L(i)$.

This problem may now be in a more convenient form for solution by dynamic programming. However, it is rarely advantageous to create new constraints except where it is recognized that a technique can be used which is well suited to handling restrictions.

## 11.4  Introducing auxiliary variables

Sometimes the whole mathematical character of a problem can be altered by introducing additional variables into a problem. We met this approach in linear programming where inequality constraints were converted into equalities by introducing slack variables; even more variables (called the artificial variables) were brought in for the determination of an initial feasible solution. These auxiliary variables did not alter the basic problem: they merely converted the problem into a convenient form for applying the simplex method. Two further examples will illustrate these possibilities.

*Example* 11.3

Suppose the objective function requires the minimization of the largest amongst a set of variables subject to a set of linear constraints on the variables. If the variables are denoted by $x_1$, $x_2$, ..., $x_N$, the objective function has the form:

minimize $\{ \max_{1 \le i \le N} x_i \}$.

This is an awkward form of function to handle. However, we can convert it neatly into a linear form by adding one new variable to the system. Let us introduce the additional variable $x_{N+1}$ and constrain the other $x_i$ values to be less than $x_{N+1}$ so that

$x_i \le x_{N+1}$ for $i = 1, ..., N$.

If we now write the objective function as

minimize $x_{N+1}$

we know that the largest $x_i$ ($1 \le i \le N$) value will equal $x_{N+1}$ as required, and, by minimizing $x_{N+1}$ we effectively minimize

$\max_{1 \le i \le N} x_i$.

The problem can now be tackled by linear programming.

*Example* 11.4

Some of the most awkward types of constraints to handle are the conditional constraints of the 'either-or' type where we require that for a pair of variables $x_j$ and $x_k$,

either

$x_j - x_k \ge a_k$

or

$x_k - x_j \ge a_j$.

If there are only a few of these constraints and otherwise the problem has a suitable form for linear programming it is possible to introduce auxiliary variables to convert them to a linear unconditional form.

Let $A$ be a number larger than $| x_j - x_k |$ will ever attain and let $y$ be an integer variable constrained to be 0 or 1. Now consider the pair of constraints

$(A + a_k)y + (x_j - x_k) \ge a_k$

$(A + a_j)(1 - y) + (x_k - x_j) \ge a_j$.

If $y = 0$ the first of the constraints corresponds to

$x_j - x_k \ge a_k$

and the second constraint states

$x_j - x_k \le A$

which is automatically satisfied. Equally if $y = 1$, the first constraint is

$$x_k - x_j \leqq A$$

and the second constraint is

$$x_k - x_j \geqq a_j.$$

The $y$ variable thus controls which constraint is active and ensures that either one or the other of the original conditional constraints must hold.

The new system of constraints includes the new 'zero-one' integer variable $y$ which is not a very attractive addition to the problem from the mathematical viewpoint, but it does mean that the problem can be solved by linear programming. It would be necessary to introduce variables like $y$ and assess quantities such as $A$ for each pair of variables which possessed these either-or type constraints; if the problem was large this could become unmanageable.

## 11.5  Transformations

Mathematical transformations have been used in a wide variety of fields for converting intractable mathematical problems into more convenient forms. In optimization studies they have been recommended for the elimination of constraints in non-linear problems where the unconstrained optimization techniques can be more powerful than the constrained procedures.

A transformation is a relationship between two sets of variables. If the original problem is defined in terms of variables

$$X = (x_1, x_2, \ldots, x_N)$$

we choose a new set of variables

$$Y = (y_1, y_2, \ldots, y_N)$$

and express a functional relationship between the two sets. The general form of the relationships will be a list of $N$ functional equations:

$$x_1 = f_1(y_1, y_2, \ldots, y_N)$$
$$x_2 = f_2(y_1, y_2, \ldots, y_N)$$
$$\vdots$$
$$x_N = f_N(y_1, y_2, \ldots, y_N)$$

and this set of equations may be summarized by the transformation $T$ where

$$X = T(Y).$$

We can now express the original objective function $F(X)$ in terms of the $y_i$ variables as

$$F'(Y) = F(X) = F\{T(Y)\}$$

and if there are $M$ constraints of the form $G_k(x) \leqq 0$ they are similarly expressed as

$$G_k'(Y) = G_k(X) = G_k\{T(Y)\} \leqq 0 \text{ for } k = 1, \ldots, M.$$

However, as the point of the transformation is generally to eliminate constraints we hope that some of these constraints will be automatically satisfied and need not be considered. We will give two examples.

*Example* 11.5

An optimization problem in three variables $x_1$, $x_2$, $x_3$ requires the minimization of

$$F(x_1, x_2, x_3) = x_1^2 + \frac{x_2}{x_3}$$

subject to the constraints

$$0 < x_1 < x_2 < x_3.$$

The problem can be transformed into the new variables $y_1$, $y_2$, $y_3$ by the relations

$$x_1 = y_1^2$$
$$x_2 = y_1^2 + y_2^2$$
$$x_3 = y_1^2 + y_2^2 + y_3^2$$

to give the new objective function

$$F'(y_1, y_2, y_3) = y_1^4 + \frac{y_1^2 + y_2^2}{y_1^2 + y_2^2 + y_3^2}.$$

However, as these constraints merely require $y_i^2 \geq 0$, $i = 1, 2, 3$ and the square of any number is positive anyway, they are automatically satisfied. The new problem therefore is to minimize

$$F'(y_1, y_2, y_3)$$

without any constraints.

*Example* 11.6

An objective function $F(x_1, x_2, x_3)$ is to be minimized subject to inequality limits on the $x_i$ values as

$$a_i \leq x_i \leq b_i.$$

We can transform this problem to eliminate the constraints by writing $x_i$ in terms of the variable $y_i$ as

$$x_i = a_i + (b_i - a_i) \sin^2 y_i.$$

Since $0 \leq \sin^2 y_i \leq 1$ for all values of $y_i$, the value of $x_i$ will lie between $a_i$ and $b_i$ when the problem is solved in terms of the $y_i$ variables. We can therefore rewrite the objective function as

$$F(a_1 + (b_1 - a_1) \sin^2 y_1, \; a_2 + (b_2 - a_2) \sin^2 y_2, \; a_3 + (b_3 - a_3) \sin^2 y_3)$$

and minimize it subject to no constraints.

However desirable it may be to use a transformation, there may be difficulties in finding a suitable one. The choice of transformations offers great scope for mathematical ingenuity. But they also introduce uncertainties. One must establish that the local optima in the transformed problem variables correspond with the local optima of the original problem. Otherwise one may be worse off than before; it would be possible for example to have a host of possible local optima for the transformed problem and a single local (and global) optimum in the original problem. The examination of the nature of the transformed surface may not be an easy task.

## 11.6 Duality in linear problems

Linear programming models possess the interesting property of forming pairs of symmetrical problems. To any maximization problem there corresponds a minimization problem involving the same data and there is a close correspondence between their optimal solutions.

The general linear programming problem with inequality constraints is to determine $x_i$ values to satisfy the system of equations

$$a_{11}x_1 + a_{12}x_2 + \ldots + a_{1N}x_N \leqq b_1$$
$$a_{21}x_1 + a_{22}x_2 + \ldots + a_{2N}x_N \leqq b_2$$
$$\cdot \quad \cdot \quad \quad \cdot \quad \quad \cdot \quad \quad \cdot$$
$$a_{M1}x_1 + a_{M2}x_2 + \ldots + a_{MN}x_N \leqq b_M$$
$$c_1x_1 + c_2x_2 + \ldots + c_Nx_N = z \text{ to be maximized}$$
$$x_i \geqq 0$$

where $a_{ij}$, $b_i$ and $c_i$ are constants, unrestricted in sign, and the objective function is written in the form of an equation in the conventional way for linear programming. Let us call this problem the primal problem. A new and apparently unrelated linear programming problem can be generated by transposing the rows and columns of this system of equations including the objective function and vector of $b_i$ values reversing the inequalities and minimizing instead of maximizing. This gives the linear programming maximization problem

$$a_{11}y_1 + a_{21}y_2 + \ldots + a_{M1}y_M \geqq c_1$$
$$a_{12}y_1 + a_{22}y_2 + \ldots + a_{M2}y_M \geqq c_2$$
$$\cdot \quad \cdot \quad \cdot \quad \cdot \quad \cdot \quad \cdot \quad \cdot \quad \cdot \quad \cdot$$
$$a_{1N}y_1 + a_{2N}y_2 + \ldots + a_{MN}y_M \geqq c_N$$
$$b_1y_1 + b_2y_2 + \ldots + b_My_M = v \text{ to be minimized}$$
$$y_i \geqq 0$$

in which the objective function is expressed as $v$ rather than $z$. We will name this problem the 'dual' problem and call the $y_i$ variables the dual variables.

There is a very close interrelationship between the solution of these two problems. The fundamental theorem of duality states that, provided a feasible solution exists, the objective function values at the optima of both problems are equivalent, i.e.

max $z$ = min $v$.

Also the solution to the primal problem automatically solves the dual and vice versa. These results are proved in an Appendix (p. 174).

The relationship between the primal and the dual is more easily recognized if the problems are written in matrix notation. For the primal, let the $M \times N$ matrix of coefficients be denoted by $A$, and let $B$ and $C$ denote the column vectors of coefficients $b_i$, $c_i$ of $M$ and $N$ dimensions respectively. If $X$ denotes a column vector of the $N$ variables $(x_1, x_2, ..., x_N)$ and $Y$ denotes a column vector of the $M$ variables $(y_1, y_2, ..., y_M)$ the primal may be written as

$A . X \leqq B$

$C' . X = z$ to be maximized

$X \geqq 0$

and the dual problem is written as

$A' . Y \geqq C$

$B' . Y = v$ to be minimized

$Y \geqq 0$

where the dashes denote transposes.

It may be noted that the dual of the dual is the primal. The dual problem may be written in the form of the primal as

$-A' . Y \leqq -C$

$-B' . Y = -v$ to be maximized

$Y \geqq 0.$

Now converting to the dual of this problem in terms of variables

$W = (w_1, w_2, ..., w_N)$

we obtain the problem

$-(A')' . W \geqq -B$

$-C' . W = z$ to be minimized

$W \geqq 0$

which may be rewritten as

$A . W \leqq B$

$C' . W = z$ to be maximized

$W \geqq 0.$

This is identical to the original primal problem.

*Example* 11.7

As an example let us take the primal problem in two variables $x_1$ and $x_2$ as

$$2x_1 + 2x_2 \leq 4$$
$$3x_2 \leq 2$$
$$4x_1 + x_2 \leq 3$$
$$5x_1 + 4x_2 = z \text{ to be maximized}$$
$$x_1, x_2 \geq 0.$$

The dual of this problem requires three variables $y_1$, $y_2$, $y_3$ and is expressed as

$$2y_1 \qquad + 4y_3 \geq 5$$
$$2y_1 + 3y_2 + y_3 \geq 4$$
$$4y_1 + 2y_2 + 3y_3 = v \text{ to be minimized}$$
$$y_1, y_2, y_3 \geq 0.$$

We will follow through the solutions to these two problems to identify their interconnections. The primal problem is solved by introducing three slack variables $x_3$, $x_4$, $x_5$ to give the system of equations

$$2x_1 + 2x_2 + x_3 \qquad = 4$$
$$3x_2 \quad + x_4 \qquad = 2$$
$$4x_1 + x_2 \qquad + x_5 = 3$$
$$5x_1 + 4x_2 \qquad = z \text{ to be maximized}$$
$$x_i \geq 0 \quad i = 1, ..., 5.$$

The optimal basis for this problem includes the variables $x_1$, $x_2$ and $x_3$ for which the canonical form is

$$x_3 - \tfrac{1}{2}x_4 - \tfrac{1}{2}x_5 = \tfrac{3}{2}$$
$$x_2 \quad + \tfrac{1}{3}x_4 \qquad = \tfrac{2}{3}$$
$$x_1 \qquad -\tfrac{1}{12}x_4 + \tfrac{1}{4}x_5 = \tfrac{7}{12}$$
$$-\tfrac{11}{12}x_4 - \tfrac{5}{4}x_5 = z - \tfrac{67}{12}.$$

The solution is $x_1 = \tfrac{7}{12}$,
$$x_2 = \tfrac{2}{3},$$
$$x_3 = \tfrac{3}{2},$$

the objective function value is $\tfrac{67}{12}$, and the relative cost coefficients for the excluded variables are

$$c_4' = -\tfrac{11}{12},$$
$$c_5' = -\tfrac{5}{4}.$$

The dual problem is solved by introducing two slack variables $y_4$ and $y_5$ to give the equations

$$2y_1 \qquad +4y_3 -y_4 \qquad = 5$$
$$2y_1 +3y_2 + y_3 \qquad -y_5 = 4$$
$$4y_1 +2y_2 +3y_3 \qquad = v \text{ to be minimized}$$
$$y_i \geq 0.$$

The optimal solution to this problem includes as basis variables $y_2$ and $y_3$ and the canonical form may be written down as

$$\tfrac{1}{2}y_1 \qquad +y_3 - \tfrac{1}{4}y_4 \qquad = \tfrac{5}{4}$$
$$\tfrac{1}{2}y_1 +y_2 \qquad +\tfrac{1}{12}y_4 -\tfrac{1}{3}y_5 = \tfrac{11}{12}$$
$$\tfrac{3}{2}y_1 \qquad +\tfrac{7}{12}y_4 +\tfrac{2}{3}y_5 = v-\tfrac{67}{12}.$$

(As this is a minimization problem we require all the relative cost coefficients to be positive at the optimum.) The solution is:

$$y_2 = \tfrac{11}{12},$$
$$y_3 = \tfrac{5}{4},$$

the objective function value is $\tfrac{67}{12}$,

and the relative cost coefficients (here written as $d_i'$) are

$$d_1' = \tfrac{3}{2},$$
$$d_4' = \tfrac{7}{12},$$
$$d_5' = \tfrac{2}{3}.$$

It will be seen that there is a close correspondence between these two solutions. The maximum value of $z$ is equal to the minimum value of $v$. Furthermore in the optimal solutions the values of the basic variables in one problem and the values of the relative cost coefficients in the other problem are closely connected. In the primal problem $c_4'$ and $c_5'$ are numerically equal to $\tfrac{11}{12}$ and $\tfrac{5}{4}$. These are the values of the solutions $y_2$, $y_3$. Also, in the dual the relative cost coefficients are $\tfrac{3}{2}$, $\tfrac{7}{12}$, $\tfrac{2}{3}$ and these are the values of the basic variables in the primal solution. The correspondence between these values is more obvious if a different notation is adopted. For the primal problem let us write the slack variables $x_3$, $x_4$, $x_5$ as $x_1^*$, $x_2^*$, $x_3^*$, and for the dual we write the two slack variables as $y_1^*$, $y_2^*$ instead of $y_4$ and $y_5$. Then the relative cost coefficients for the excluded variables in the primal relate to the variables $x_2^*$ and $x_3^*$ with numerical values $\tfrac{11}{12}$ and $\tfrac{5}{4}$. Furthermore, if we define 'corresponding' variables as $x_i^*$ and $y_i$, the basic variables in the dual correspond to the non-basic variables in the optimal solution to the primal. The basic variables in the dual solution corresponding to the variables $x_2^*$ and $x_3^*$ are $y_2$ and $y_3$ with values $\tfrac{11}{12}$ and $\tfrac{5}{4}$. The point to note is that the suffices of the

$x^*$ correspond to the suffices of the $y$. Similarly, the other way round, the non-basic variables in the dual are

$y_1, y_1^*, y_2^*$

with relative cost coefficients $\frac{3}{2}$, $\frac{7}{12}$ and $\frac{2}{3}$ and these correspond to the basic variables in the solution of the primal, i.e.

$x_1^*, x_1, x_2.$

We will use this convenient notation for the proof of duality given in the Appendix (p. 174).

The connection between the primal and the dual problems means that if a solution to either problem is obtained then the other problem is automatically solved. Sometimes this option can be exploited to a considerable advantage for computational purposes. For instance if we have a four variable problem with two inequality constraints we can solve it graphically by creating the dual problem which will only have two structural variables and four constraints.

## 11.7 Interpretation of the dual problem

The duality phenomenon can provide useful and interesting interpretations about a problem. The dual variable values which are available from the linear programming solution to the primal problem provide information about the effect of changing any of the constraint limitations. Suppose the primal objective function to be maximized is expressed as

$$z = \sum_{i=1}^{N} c_i x_i$$

and the dual objective function to be minimized is

$$v = \sum_{i=1}^{M} b_i y_i.$$

By the duality theorem, we have at the optimum

$$\max z = \min v = \sum_{i=1}^{M} b_i \bar{y}_i$$

where $\bar{y}_i$ are the optimal solutions. Now a typical constraint in the primal is

$$\sum_{j=1}^{N} a_{ij} x_j \leqq b_i$$

so that, at the optimal basis, the $\bar{y}_i$ quantities measure the effect on $z$ of a unit change in $b_i$.

Consider the problem of a firm making a number of different products, each of which is a mixture of a certain number of resources and desiring to make a maximum profit. Let there be $N$ products and $M$ raw materials and let $a_{ij}$ denote the proportional input of material $i$ into product $j$. Then if $c_j$ is the return from selling 1 unit of product $j$ and $b_i$ is the limited supply of raw material, the problem is to

$$\text{maximize } \sum_{j=1}^{N} c_j x_j = z$$

subject to

$$\sum_{j=1}^{N} a_{ij} x_j \leqq b_i \text{ for } i = 1, \ldots, M.$$

Now suppose that the firm has the opportunity to increase or decrease resource $i$. As before

$$\max z = \sum_{j=1}^{N} c_j \bar{x}_j = \sum_{i=1}^{M} b_i \bar{y}_i = \min v$$

where $\bar{x}_j$ and $\bar{y}_i$ are the optimal solutions. This shows that the value of the maximum profit would increase or decrease by an amount $\bar{y}_i$ times the alteration to the resource level $b_i$. The dual variables are often called the marginal or shadow prices or imputed costs of the resources. It should be noted that if constraint $i$ is not binding, i.e.

$$\sum_{j=1}^{N} a_{ij} \bar{x}_j < b_i$$

then the slack variable $x_i^*$ will be in the basis and the corresponding dual variable $y_i$ will not be in the dual basis and has value 0. This implies that the shadow price is zero and it is not worth altering the level of this resource. The result agrees with commonsense as surplus or spare resources are available.

Duality is a concept which is not restricted to linear programming. It also arises in non-linear optimization and in other areas of mathematics, economics, engineering, physics and many other fields. For example in electric circuit theory, duality exists between series and parallel circuits. The same form of equation describes both circuits and it is simply a matter of interpretation whether we are dealing with a voltage or current generator. A quadratic duality theorem has been applied to a problem in applied mechanics in the analysis of elastic-plastic structures. Here the dual problem produced a generalization of the principle of minimum potential energy which, as far as is known, was not discovered till the duality theorem produced it. The main achievements of duality have been in the field of problem interpretation. Apart from its connection with the revised simplex method in linear programming the ideas of duality have not yet contributed significantly to successful optimization techniques.

# Appendix

## *11.8 Proof of the duality theorem

The linear programming problem in $N$ variables with $M$ inequalities is

$A.X \leq B$

$C'X = z$ to be maximized

$X \geq 0$

where $A$ is the $M \times N$ matrix of coefficients $a_{ij}$, $X$ and $B$ the column vectors of the $x_j$ variables and $b_i$ quantities respectively, and $C'$ is the row vector of the $c_j$ coefficients. This problem has a dual with $N$ inequalities in $M$ variables:

$A'.Y \geq C$, $Y \geq 0$, $B'Y = v$ to be minimized

where $A'$ is the transpose of $A$, and $Y$ is a column vector of variables $y_i$. The two problems can be rewritten with slack variables as

$AX + I_M X^* = B$, $X \geq 0$, $X^* \geq 0$, $C'X = z$ to be minimized

$A^* Y - I_N Y^* = C$, $Y \geq 0$, $Y^* \geq 0$, $B'Y = v$ to be maximized,

where $I_M$ and $I_N$ are identity matrices of dimensions $M$ and $N$, and $X^*$ and $Y^*$ are column vectors of slack variables $x_1^*$, $x_2^*$, ..., $x_M^*$ and $y_1^*$, $y_2^*$, ..., $y_N^*$ respectively.

We will refer to the first problems in the $X$ variables as the primal problem and call the second problem in the $Y$ variables the dual problem. The correspondence ('duality') between the solutions to the two problems is stated by the following theorem. (We use the notation developed in section 11.6.)

*Duality theorem*: In the optimal basic solutions of the primal and dual problems, the values of $x_j$ and $x_i^*$ in the basis are numerically equal to the relative cost coefficients of $y_j^*$ and $y_i$, and the values of $y_i$ and $y_j^*$ in the basis are numerically equal to the relative cost coefficients of $x_i^*$ and $x_j$ respectively, and

$\max z = \min v$.

*Proof*: We will assume that the optimal basis of the primal consists of original variables only and includes no slack variables. We can then partition the vector $X$ into the vector of $M$ basic variables, $X_1$, and the vector of the remaining original variables, $X_2$; let $A$ and $C'$ be partitioned correspondingly and the primal can then be written as

$A_1 X_1 + A_2 X_2 + I_M X^* = B$

$C_1 X_1 + C_2 X_2 = z$ to be maximized

$X_1 \geq 0$, $X_2 \geq 0$, $X^* \geq 0$.

Solving the equation system for $X_1$ we get

$$X_1 = A_1^{-1} \cdot B - A_1^{-1} \cdot A_2 X_2 - A_1^{-1} X^*$$

and substituting in the equation for $z$ gives

$$z = C_1' A_1^{-1} B - (C_1' A_1^{-1} \cdot A_2 - C_2') X_2 - C_1' A_1^{-1} X^*.$$

Since we have assumed that $X_1$ is the optimal basis, these equations give the optimal solution for $X_2 = X^* = 0$.

Now the dual problem can be written, after partitioning $I_N$ and $Y^*$, as

$$A_1' Y - I_M Y_1^* \qquad\qquad = C_1$$

$$A_2' Y \qquad\quad - I_{N-M} Y_2^* = C_2$$

$B'Y = v$ to be minimized

$$Y \geqq 0, \; Y_1^* \geqq 0, \; Y_2^* \geqq 0.$$

If we take $(Y, Y_2^*)$ as a basis for the solution of the dual we get

$$Y = (A_1')^{-1} C_1 + (A_1')^{-1} Y_1^*$$

$$Y_2^* = [A_2'(A_1')^{-1} C_1 - C_2] + A_2'(A_1')^{-1} Y_1^*$$

and substitution in $v$ gives

$$v = B'(A_1')^{-1} C_1 + B'(A_1')^{-1} Y_1^*.$$

The basic solution for $Y$ and $Y_2^*$ is non-negative since it is equal to the (transposed) relative cost coefficients of $X^*$ and $X_2$ in the expression for $z$. It is optimal because the simplex coefficient of $Y_i^*$ are positive, being equal to the (transposed) basic solution for $X_1$. Furthermore, the constant terms in the expressions for $z$ and $v$ are the transposes of each other, i.e.

$$\max z = \min v.$$

The proof is now complete.

The proof of the case where slack variables enter the optimal basis of the primal follows along similar lines and can be found in texts on linear programming.

REFERENCES

Box, M. J. 1966. A comparison of several unconstrained optimization methods and the use of transformations in constrained problems. *Computer J.*, 9, 67.
Dano, S. 1960. *Linear Programming in Industry—Theory and Applications.* Springer-Verlag.
Gass, S. I. 1958. *Linear Programming, Methods and Applications.* McGraw-Hill, New York.
Kaufmann, A. 1963. *Methods and Models of Operations Research.* Prentice-Hall, New York.

## Exercises on Chapter 11

**1** A minimization problem in one variable has an objective function

$$F(x) = x^2 - 4x + 4$$

and a single constraint

$$x \geq 1.$$

Sketch the forms of the revised objective functions if the constraint is embodied in the objective function by the two methods described in Example 11.1. Set the parameter $r$ to $\frac{1}{10}$ for the response surface method and to 10 for the alternative method.

**2** Express the following problem in the linear programming standard form:

Maximize $(\min (x_1, x_2, x_3))$

subject to $2x_1 + 3x_2 + 6x_3 \leq 12$

$\qquad\qquad 4x_1 + 2x_2 + 5x_3 \leq 15$

$\qquad\qquad x_1, x_2, x_3 \geq 0.$

**3** Convert the following problem with two pairs of conditional constraints into an integer linear form:

Maximize $5x_1 + 3x_2 + x_3$

subject to $x_1 - x_3 \leq 5$

and

either $x_1 - x_2 \geq 12$

or $\qquad x_2 - x_1 \geq 8.$

$x_1, x_2, x_3$ are constrained to lie in the ranges

$0 \leq x_1 \leq 20$

$0 \leq x_2 \leq 15$

$0 \leq x_3 \leq 10.$

**4** Transform the following constrained problem in continuous variables $x_1, x_2, x_3$ into an unconstrained form.

Maximize $-x_1^2 + 3(x_2)^{\frac{1}{2}}(x_1 + x_3)$

subject to $\qquad x_1 \geq 0$

$\qquad\qquad 0 \leq x_2 \leq 5$

and

$\qquad\qquad -4 \leq x_3 \leq 6.$

**5** Given the primal linear programming problem in $x_1$, $x_2$

$2x_1 + 3x_2 \leqq 1$

$x_1 + 7x_2 \leqq 1$

$8x_1 + 21x_2 = z$ to be maximized

$x_1 \geqq 0$

$x_2 \geqq 0,$

express the dual problem, and confirm that the objective functions have the same values at the optimum.

**6** Find the minimum value of the function

$12x_1 + 24x_2 + 15x_3$

subject to

$x_1 + 4x_2 + 3x_3 \geqq 2$

$2x_1 + 3x_2 + x_3 \geqq 1$

$x_i \geqq 0 \quad i = 1, 2, 3,$

by a graphical method.

**7** A blending problem requires two quantities $x_1$, $x_2$ to be mixed so that the mixture will contain the prescribed minimum amounts $b_1$, $b_2$ of two constituents. These conditions are expressed by linear conditions

$a_{11}x_1 + a_{12}x_2 \geqq b_1$

$a_{21}x_1 + a_{22}x_2 \geqq b_2.$

The cost of mixing amounts $x_1$ and $x_2$ is $c_1x_1 + c_2x_2$. The objective is to minimize the costs. Interpret the dual to this problem.

**\*8** The method of separable programming requires that cross product terms such as $x_1x_2$ do not occur in the objective or constraint functions. By writing

$y_1 = x_1 + x_2,$

$y_2 = x_1 - x_2,$

show how the cross product term can be re-expressed as the sum of separate functions.

**\*9** Some of the techniques for integer optimization problems restrict the integer variables to the values 0 or 1. Show that an integer variable which could lie between 0 and some positive integer $v$ can be expressed in terms of a sum of zero-one variables each multiplied by a power of 2.

# Answers to exercises

**Chapter 2** (p. 15)

**1**

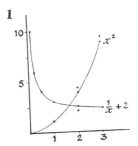

Fig. A2.1

**2** $\dfrac{df}{dx} = \dfrac{3}{2\sqrt{x}} \left( \sqrt{x} - \dfrac{1}{\sqrt{x}} \right)^2 \left( \dfrac{x+1}{x} \right)$

$\dfrac{\partial F}{\partial x_1} = \dfrac{-x_2 e^{-x_1}((1-x_1)\log(x_1 x_2) - x_1)}{(1 + \log x_1 x_2)^2}$

$\dfrac{\partial F}{\partial x_2} = \dfrac{x_1 e^{-x_1} \log x_1 x_2}{(1 + \log x_1 x_2)^2}.$

**3**

| $x$ | $f(x)$ | $f(1)+(x-1)\left[\dfrac{df}{dx}\right]_{x=1}$ | $f(1)+(x-1)\left[\dfrac{df}{dx}\right]_{x=1} +\dfrac{(x-1)^2}{2}\left[\dfrac{d^2 f}{dx^2}\right]_{x=1}$ |
|---|---|---|---|
| 1·1 | 1·331 | 1·30 | 1·33 |
| 1·25 | 1·95 | 1·75 | 1·94 |
| 1·50 | 3·38 | 2·5 | 3·25 |
| 2·0 | 8·0 | 4·0 | 7·0 |

**4** $10 - e^{-3}$.

**5** $X.Y = 6$. The length of $X$ is $\sqrt{60}$ and of $Y$ is $\sqrt{18}$.

**6** For the vectors to be linearly dependent we require that quantities $a_1$, $a_2$, $a_3$ can be chosen, not all zero such that

$$a_1 + 3a_2 + 2a_3 = 0$$
$$2a_1 - a_2 \qquad = 0$$
$$2a_1 \qquad + 3a_3 = 0.$$

If $a_1 \neq 0$, by the second and third equations,

$$\frac{a_2}{a_1} = 2 \text{ and } \frac{a_3}{a_1} = -\tfrac{2}{3}.$$

Substituting in the first equation after dividing through by $a_1$, $(1 + 6 - \tfrac{4}{3})$ should be zero. This is impossible, hence $a_1 = 0$. But if $a_1 = 0$, $a_2 = 0$ by the second equation, and then $a_3 = 0$ by the third equation. It is therefore impossible to find any quantities $a_1$, $a_2$, $a_3$ not all zero and the vectors must be linearly independent.

**7** The relevant region is shown shaded.

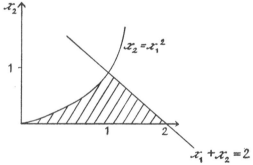

Fig. A2.2

**8** The point of intersection of the line and the curve is (1, 1). The diagram of Fig. A2.2 suggests choosing the points (0, 0) and (1, 1) as two points in the region. The point halfway along the chord joining these two points has coordinates $(\tfrac{1}{2}, \tfrac{1}{2})$. This point must be inside the region for the region to be convex.

But, putting this point in the first constraint, we require $-\tfrac{1}{4} + \tfrac{1}{2} \leqq 0$ is clearly false.

Therefore, the point $(\tfrac{1}{2}, \tfrac{1}{2})$ does not lie in the region, and the region is not convex.

**9** (i) $(A + B) = \begin{pmatrix} 2 & 3 & 4 \\ 2 & 5 & 9 \end{pmatrix}$, $A.B$ meaningless, $(A+B)' = \begin{pmatrix} 2 & 2 \\ 3 & 5 \\ 4 & 9 \end{pmatrix}$.

G

(ii) $(A+B)$ meaningless, $A \cdot B = \begin{pmatrix} 1 & 3 & 5 & 7 \\ 3 & 7 & 11 & 15 \\ 5 & 11 & 17 & 23 \\ 7 & 15 & 23 & 31 \end{pmatrix}$, $(A \cdot B)' = A \cdot B$.

**10** $\begin{pmatrix} 3 & 5 \\ 8 & 10 \end{pmatrix}\begin{pmatrix} b_{11} & b_{12} \\ b_{21} & b_{22} \end{pmatrix} = \begin{pmatrix} 1 & 0 \\ 0 & 1 \end{pmatrix}$ gives the four equations

$\left.\begin{array}{l} 3b_{11}+5b_{21} = 1 \\ 8b_{11}+10b_{21} = 0 \end{array}\right\}$ giving $b_{11} = -1$, $b_{21} = 0 \cdot 8$

$\left.\begin{array}{l} 3b_{12}+5b_{22} = 0 \\ 8b_{12}+10b_{22} = 1 \end{array}\right\}$ giving $b_{12} = 0 \cdot 5$, $b_{22} = -0 \cdot 3$.

Therefore the inverse is $\begin{pmatrix} -1 & 0 \cdot 5 \\ 0 \cdot 8 & -0 \cdot 3 \end{pmatrix}$.

**11** There are five permutations which can be obtained by making adjacent exchanges on the permutation $(e_6, e_1, e_4, e_3, e_2, e_5)$, they are:

$e_1, e_6, e_4, e_3, e_2, e_5$

$e_6, e_4, e_1, e_3, e_2, e_5$

$e_6, e_1, e_3, e_4, e_2, e_5$

$e_6, e_1, e_4, e_2, e_3, e_5$

$e_6, e_1, e_4, e_3, e_5, e_2$.

**12** The connections are $(1, 2)$, $(1, 3)$, $(1, 5)$, $(2, 3)$, $(2, 4)$, $(2, 5)$, $(3, 4)$, $(4, 5)$. The network may be drawn up as shown in Fig. A2.3.

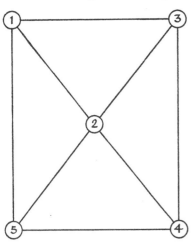

Fig. A2.3

**13** (*i*) Mean is 2, Variance is 28/15.

   (*ii*) Mean is ½, Variance is 1/20.

Expected value is $2\left(1 - \dfrac{2}{e}\right)$.

## Chapter 3 (p. 27)

**1** The table gives the relevant values of $F(x_1, x_2)$

**Table A3.1**

|  | $x_1$ 0 | 1 | 2 | 3 | 4 |
|---|---|---|---|---|---|
| 0 | 10 | 6 | 4 | 4 | 6 |
| 1 | 4 | 0 | −2 | −2 | 0 |
| 2 | 0 | −4 | −6 | −6 | −4 |
| 3 | −2 | −6 | −8 | −8 | −6 |
| 4 | −2 | −6 | −8 | −8 | −6 |
| 5 | 0 | −4 | −6 | −6 | −4 |

($x_2$ labels the rows.)

The minimum lies at any of the points (2, 3), (2, 4), (3, 3), (3, 4). The maximum occurs at the point (0, 0).

**2** The feasible region is shown shaded in Fig. A3.1.

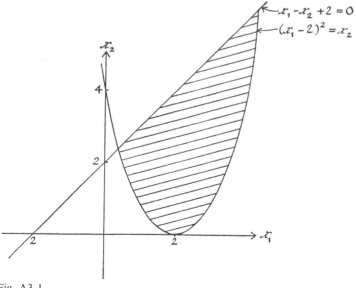

Fig. A3.1

**3** The feasible region is shown shaded in Fig. A3.2.

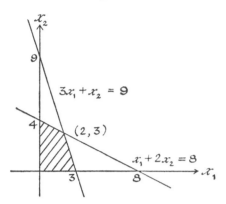

Fig. A3.2

**4** The feasible region consists of the 8 points marked in Fig. A3.3.

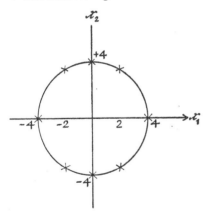

Fig. A3.3

**5**

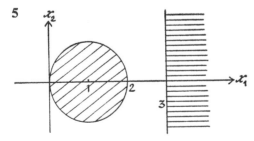

Fig. A3.4

A feasible point must be in one or other of the shaded areas in Fig. A3.4.

6

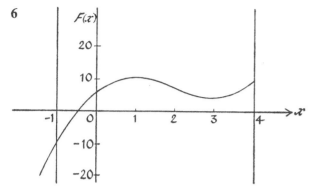

Fig. A3.5

Local minima $x = -1, 3$, global minimum at $x = -1$.
Local maxima $x = 1, 4$, both are also global.

7

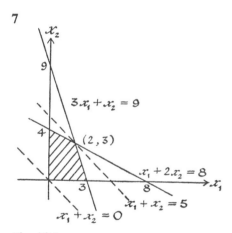

Fig. A3.6

(*i*) Feasible region is specified for all values of $a \leq 5$.

(*ii*) The single point is (2, 3) for $a = 5$.

(*iii*) Feasible region is non-existent for $a > 5$.

## Chapter 4 (p. 40)

**1** The derivative is discontinuous at $x = 0$.

**2** By calculus, the minimum occurs where $x = 1$, and this violates the constraint $x \geq 2$.

**3** From the derivative equation and consideration of the boundary points, local maxima occur at $x = 4, 9$, local minima at $x = 0, 8$.

**4** $\dfrac{dF}{dx} = 3x^2 = 0$ for $x = 0$. Also $\dfrac{d^2F}{dx^2} = 6x = 0$ for $x = 0$. But $3x^2 \geq 0$ for $x > 0$ and $x < 0$. Hence it is a point of inflexion.

**5** $\dfrac{dF}{dx} = \dfrac{ad - cb}{(cx + d)^2} \geq 0$ for all values of $x$ if $(ad - cb) > 0$

$\qquad\qquad \leq 0$ for all values of $x$ if $(ad - cb) < 0$

$\qquad\qquad = 0$ if $ad = cb$, for all $x$ except $x = -\dfrac{d}{c}$ where it is undefined.

Therefore the derivative is either always zero or always has the same sign. Therefore there are no maxima or minima.

**6** (*i*) At the origin:

$$\frac{\partial F}{\partial x_1} = 2x_1 + x_2 + 3x_1^2 + 2x_1x_2 \quad = 0$$

$$\frac{\partial^2 F}{\partial x_1^2} = 2 + 6x_1 + 2x_2 \qquad\qquad = 2$$

$$\frac{\partial^2 F}{\partial x_1 \partial x_2} = 1 + 2x_1 \qquad\qquad = 1$$

$$\frac{\partial F}{\partial x_2} = x_1 + 2x_2 + x_1^2 + 3x_2^2 \qquad = 0$$

$$\frac{\partial^2 F}{\partial x_1^2} = 2 + 6x_2 \qquad\qquad = 2$$

$$\frac{\partial^2 F}{\partial x_1^2}\frac{\partial^2 F}{\partial x_2^2} - \left(\frac{\partial^2 F}{\partial x_1 \partial x_2}\right)^2 = 3 > 0$$

$$\frac{\partial^2 F}{\partial x_1^2} \qquad\qquad\qquad > 0$$

This is a local minimum.

(*ii*) At the origin:

$$\frac{\partial F}{\partial x_1} = 2x_1 + 3x_2 + 3x_1^2 + 2x_1x_2 = 0$$

$$\frac{\partial^2 F}{\partial x_1^2} = 2 + 6x_1 + 2x_2 \qquad\qquad = 2$$

$$\frac{\partial^2 F}{\partial x_1 \partial x_2} = 3 + 2x_1 \qquad = 3$$

$$\frac{\partial F}{\partial x_2} = 3x_1 + 2x_2 + x_1^2 + 3x_2^2 \qquad = 0$$

$$\frac{\partial^2 F}{\partial x_2^2} = 2 + 6x_2 \qquad = 2$$

$$\frac{\partial^2 F}{\partial x_1^2} \frac{\partial^2 F}{\partial x_2^2} - \left(\frac{\partial^2 F}{\partial x_1 \partial x_2}\right)^2 = -5 < 0 \text{ giving a saddle point.}$$

(*iii*) At the origin:

$$\frac{\partial F}{\partial x_1} = 2x_1 + 2x_2 + 4x_1^3 + 2x_1 x_2^2 = 0$$

$$\frac{\partial^2 F}{\partial x_1^2} = 2 + 12x_1^2 + 2x_2^2 \qquad = 2$$

$$\frac{\partial^2 F}{\partial x_1 \partial x_2} = 2 + 4x_1 x_2 \qquad = 2$$

$$\frac{\partial F}{\partial x_2} = 2x_1 + 2x_2 + 2x_1^2 x_2 + 4x_2^3 = 0$$

$$\frac{\partial^2 F}{\partial x_2^2} = 2 + 2x_1^2 + 12x_2^2 \qquad = 2$$

$$\frac{\partial^2 F}{\partial x_1^2} \frac{\partial^2 F}{\partial x_2^2} - \left(\frac{\partial^2 F}{\partial x_1 \partial x_2}\right)^2 = 0 \text{ which is indeterminate.}$$

**7** The distance from the point $(x_1, x_2, x_3)$ is
$$F(x_1, x_2, x_3) = \{(x_1 - 1)^2 + (x_3 - 2)^2 + (x_3 - 3)^2\}^{\frac{1}{2}}.$$

However, it is simpler and equivalent to deal with the square of the function. Therefore we want the maximum value of $(F(x_1, x_2, x_3))^2$ subject to the constraint

$$x_1^2 + x_2^2 + x_3^2 = 1.$$

Introducing the Lagrange multiplier $y$ we differentiate

$$H(x_1, x_2, x_3, y) = (F(x_1, x_2, x_3))^2 + y(x_1^2 + x_2^2 + x_3^2 - 1), \text{ to give:}$$
$$2(x_1 - 1) + 2yx_1 = 0$$
$$2(x_1 - 2) + 2yx_2 = 0$$
$$2(x_3 - 3) + 2yx_3 = 0$$

and the constraint equation $x_1^2 + x_2^2 + x_3^2 = 1$.

Hence $x_1 = \dfrac{1}{y+1}$, $x_2 = \dfrac{2}{y+1}$, $x_3 = \dfrac{3}{y+1}$.

Inserting these values in the constraint equation

$14 = (y+1)^2$,

or $y^2 + 2y - 13 = 0$

giving $y = -1 \pm \sqrt{14}$.

Therefore for a maximum, by inspection of the function $F(x_1, x_2, x_3)$,

$$x_1 = -\frac{1}{\sqrt{14}}, \ x_2 = -\frac{2}{\sqrt{14}}, \ x_3 = -\frac{3}{\sqrt{14}}.$$

**8** The ellipse lies symmetrically around the origin cutting the $x_1$-axis at $x_1 = \pm a$ and the $x_2$-axis at $x_2 = \pm b$. The vertices of the rectangle will clearly lie on the ellipse. Let $(x_1, x_2)$ be the corner of the rectangle in the first quadrant. Then we wish to maximize $4(x_1 + x_2)$ subject to

$$\frac{x_1^2}{a^2} + \frac{x_2^2}{b^2} = 1.$$

Introducing a Lagrange multiplier and differentiating we obtain:

$$4 + y\frac{2x_1}{a^2} = 0$$

$$4 + y\frac{2x_2}{b^2} = 0.$$

Hence $x_1 = -\dfrac{2a^2}{y}$, $x_2 = -\dfrac{2b^2}{y}$.

Substituting in the constraint equation

$$1 = \frac{4a^2}{y^2} + \frac{4b^2}{y^2}$$

i.e. $y = \pm 2(a^2 + b^2)^{\frac{1}{2}}$.

Hence $x_1 = \dfrac{a^2}{(a^2 + b^2)^{\frac{1}{2}}}$, $x_2 = \dfrac{b^2}{(a^2 + b^2)^{\frac{1}{2}}}$.

**9** $\dfrac{\partial F}{\partial x_1} = x_2 - \dfrac{1}{x_1^2} = 0$

$\dfrac{\partial F}{\partial x_2} = x_1 - \dfrac{2}{x_2^2} = 0.$

At the optimum, $x_1 = (\tfrac{1}{2})^{\frac{1}{3}}$, $x_2 = (2)^{\frac{1}{3}}$.

$$\frac{\partial^2 F}{\partial x_1^2} = \frac{2}{x_1^3} = 4$$

$$\frac{\partial^2 F}{\partial x_1 \partial x_2} = 1$$

$$\frac{\partial^2 F}{\partial x_2^2} = \frac{4}{x_2^3} = 1$$

$$\frac{\partial^2 F}{\partial x_1^2}\frac{\partial^2 F}{\partial x_2^2} - \left(\frac{\partial^2 F}{\partial x_1 \partial x_2}\right)^2 = 3 > 0.$$

Hence the point $((\frac{1}{2})^{\frac{1}{3}}, (2)^{\frac{1}{3}})$ is a minimum.

**10** $\dfrac{\partial F}{\partial x_1} = 2x_1 e^{-(x_1^2+x_2^2)}(a - ax_1^2 - bx_2^2)$

$\dfrac{\partial F}{\partial x_2} = 2x_2 e^{-(x_1^2+x_2^2)}(b - ax_1^2 - bx_2^2).$

These equations are zero when the following conditions hold:

$x_1 = x_2 = 0$ giving $F = 0$, which is a minimum;

$x_1 = 0, x_2 = \pm 1$ giving $F = be^{-1}$, which is a maximum as $a < b$;

$x_1 = \pm 1, x_2 = 0$ giving $F = ae^{-1}$, a local maximum.

**11** $H(X, Y) = a_1 x_1^2 + a_2 x_2^2 + a_3 x_3^2 + y_1(x_1^2 + x_2^2 + x_3^2 - 1)$
$$+ y_2(b_1 x_1 + b_2 x_2 + b_3 x_3).$$

$\dfrac{\partial H}{\partial x_1} = 2a_1 x_1 + 2y_1 x_1 + y_2 b_1 = 0,$

$\dfrac{\partial H}{\partial x_2} = 2a_2 x_2 + 2y_1 x_2 + y_2 b_2 = 0,$

$\dfrac{\partial L}{\partial x_3} = 2a_3 x_3 + 2y_1 x_3 + y_2 b_3 = 0.$

Multiplying these three equations by $x_1, x_2, x_3$ respectively and using the constraints,

$$a_1 x_1^2 + a_2 x_2^2 + a_3 x_3^2 = -y_1.$$

Therefore to determine the maximum of the original function we merely need to determine $y_1$. From the three equations

$$x_1 = \frac{-y_2 b_1}{2(a_1 + y_1)}, \quad x_2 = \frac{-y_2 b_2}{2(a_2 + y_1)}, \quad x_3 = \frac{-y_2 b_3}{2(a_3 + y_1)}.$$

These values can be substituted in $b_1 x_1 + b_2 x_2 + b_3 x_1 = 0$, to give

$$\frac{b_1^2}{a_1 + y_1} + \frac{b_2^2}{a_2 + y_1} + \frac{b_3^2}{a_3 + y_1} = 0.$$

This is a quadratic in $y_1$. The least root will determine the maximum of the original objective function.

**12** Let the edges be $x_1, x_2, x_3$ where $x_1 \geq x_2 \geq x_3$.
The constraints are

$$4(x_1 + x_2 + x_3) = a$$

$$2(x_1 x_2 + x_2 x_3 + x_3 x_1) = \frac{a^2}{25}.$$

We wish to maximize

$$x_1 x_2 x_3 - x_3^3.$$

$$H(X, M) = x_1 x_2 x_3 - x_3^3 + y_1(x_1 + x_2 + x_3 - a/4) + y_2(x_1 x_2 + x_2 x_3 + x_3 x_1 - a^2/50)$$

$$\frac{\partial H}{\partial x_1} = x_2 x_3 + y_1 + y_2(x_2 + x_3) = 0,$$

$$\frac{\partial H}{\partial x_2} = x_3 x_1 + y_1 + y_2(x_3 + x_1) = 0,$$

$$\frac{\partial H}{\partial x_3} = x_1 x_2 - 3x_3^2 + y_1 + y_2(x_1 + x_2) = 0.$$

From the first two equations, whatever the values of $x_3$, $y_1$ and $y_2$, $x_1 = x_2$. Substituting in the constraint equations

$$2x_1 + x_3 \qquad\qquad = a/4$$

$$(2x_1^2 + 2x_1 x_3) \qquad\quad = a^2/25$$

giving $x_1 = x_2 = \dfrac{a}{10}$, $x_3 = \dfrac{a}{20}$.

**Chapter 5** (p. 64)
**1** Multiply the objective function by $-1$ to make it a maximization problem.
Set $x_3 = x_3^+ = x_3^-$

$$x_4 = x_4^+ - x_4^-$$

and introduce slack variables $x_5$, $x_6$. Also multiply the last equality constraint by $-1$. Then the problem is

maximize $3x_1 + 4x_2 - 2x_3^+ + 2x_3^- + x_4^+ - x_4^-$
subject to the constraints

$$3x_1 + x_2 + x_3^+ - x_3^- + x_5 \quad\quad = 7$$

$$4x_1 + x_2 - 6x_3^+ + 6x_3^- \quad\quad -x_6 = 6$$

$$x_1 + x_2 - x_3^+ + x_3^- - x_4^+ + x_4^- = 4$$

and $x_1, x_2, x_3^+, x_3^-, x_4^+, x_4^-, x_5, x_6 \geqq 0.$

2 (a)

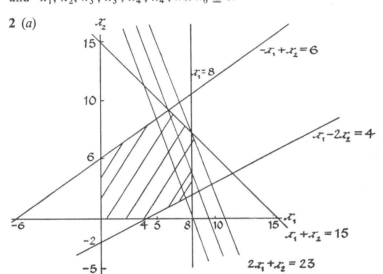

Fig. A5.1

Optimum solution is at $x_1 = 8$, $x_2 = 7$ and $z = 23$.

(b) In standard form the problem is

$$
\begin{aligned}
x_1 - 2x_2 + x_3 &= 4 \\
x_1 + \quad\quad + x_4 &= 8 \\
x_1 + x_2 \quad + x_5 &= 15 \\
-x_1 + x_2 \quad\quad + x_6 &= 6 \\
2x_1 + x_2 &= z.
\end{aligned}
$$

Initial basis $x_3 = 4$, $x_4 = 8$, $x_5 = 15$, $x_6 = 6$.

Objective function value zero.

Introduce $x_1$ exclude $x_3$, use equation 1 to obtain new canonical form

$$
\begin{aligned}
x_1 - 2x_2 + x_3 &= 4 \\
2x_2 - x_3 + x_4 &= 4 \\
3x_2 - x_3 + x_5 &= 11 \\
-x_2 + x_3 + x_6 &= 10 \\
5x_2 - 2x_3 &= z - 8.
\end{aligned}
$$

New basis is $x_1 = 4$, $x_4 = 4$, $x_5 = 11$, $x_6 = 10$.

Objective function value is 8.

Introduce $x_2$, exclude $x_4$, use equation 2 to obtain new canonical form:

$$
\begin{aligned}
x_1 \qquad\quad + x_4 \qquad\qquad\qquad &= 8 \\
x_2 - \tfrac{1}{2}x_3 + \tfrac{1}{2}x_4 \qquad\qquad\quad &= 2 \\
\tfrac{1}{2}x_3 - \tfrac{3}{2}x_4 + x_5 \qquad\quad &= 5 \\
\tfrac{1}{2}x_3 + \tfrac{1}{2}x_4 \qquad + x_6 &= 12 \\
\tfrac{1}{2}x_3 - \tfrac{5}{2}x_4 \qquad\qquad &= z - 18.
\end{aligned}
$$

New basis is $x_1 = 8$, $x_2 = 2$, $x_5 = 5$, $x_6 = 12$.

Objective function value is $-18$.

Introduce $x_3$, exclude $x_5$, use equation 3 to obtain new form:

$$
\begin{aligned}
x_1 \qquad\quad + x_4 \qquad\qquad\quad &= 8 \\
x_2 \qquad - x_4 + x_5 \qquad\quad &= 7 \\
x_3 - 3x_4 + 2x_5 \qquad\quad &= 10 \\
2x_4 - x_5 + x_6 &= 7 \\
- x_4 - x_5 \qquad\quad &= z - 23.
\end{aligned}
$$

New basis is $x_1 = 8$, $x_2 = 7$, $x_3 = 10$, $x_6 = 7$ and objective function is 23.
This is optimal.

**3** Including slack and artificial variables the problem is

$$
\begin{aligned}
x_1 + 3x_2 + x_3 \qquad\qquad\qquad\qquad &= 9 \\
x_1 + x_2 \qquad + x_4 \qquad\qquad\qquad &= 6 \\
x_1 - x_2 \qquad\qquad - x_5 \qquad + x_7 \qquad &= 2 \\
x_1 + x_2 \qquad\qquad\qquad - x_6 \qquad + x_8 &= 3 \\
\text{Maximize } 2x_1 \qquad\quad - x_5 - x_6 \qquad\qquad &= w + 5
\end{aligned}
$$

which is the infeasibility form.

The initial basis is $x_3 = 9$, $x_4 = 6$, $x_7 = 2$, $x_8 = 3$.

The objective function value is $-5$.

Introduce $x_1$, exclude $x_7$, and use the third equation to obtain the new form

$$
\begin{aligned}
4x_2 + x_3 \qquad + x_5 \qquad - x_7 \qquad\qquad &= 7 \\
2x_2 \qquad + x_4 + x_5 \qquad - x_7 \qquad\qquad &= 4 \\
x_1 - x_2 \qquad\qquad - x_5 \qquad + x_7 \qquad\qquad &= 2 \\
2x_2 \qquad\qquad + x_5 - x_6 - x_7 + x_8 &= 1 \\
2x_2 \qquad\qquad + x_5 - x_6 - 2x_7 \qquad &= w + 1.
\end{aligned}
$$

The artificial objective function is now $-1$.

Introduce $x_2$, exclude $x_8$, and use the fourth equation to obtain the new form:

$$x_3 \quad - x_5 + 2x_6 + x_7 - 2x_8 = 5$$
$$x_4 \quad + x_6 \quad - x_8 = 3$$
$$x_1 \quad -\tfrac{1}{2}x_5 - \tfrac{1}{2}x_6 - \tfrac{1}{2}x_7 + \tfrac{1}{2}x_8 = \tfrac{5}{2}$$
$$x_2 \quad +\tfrac{1}{2}x_5 - \tfrac{1}{2}x_6 - \tfrac{1}{2}x_7 + \tfrac{1}{2}x_8 = \tfrac{1}{2}$$
$$- x_7 - x_8 = w.$$

The value of $w$ is now 0 and an initial feasible basis has been obtained giving the solution

$x_1 = \tfrac{5}{2}, x_2 = \tfrac{1}{2}, x_3 = 5, x_4 = 3.$

The solution is displayed in the following tableaux:

| Iteration | Basis | Values | $x_1$ | $x_2$ | $x_3$ | $x_4$ | $x_5$ | $x_6$ | $x_7$ | $x_8$ |
|---|---|---|---|---|---|---|---|---|---|---|
| 1 | $x_3$ | 9 | 1 | 3 | 1 | | | | | |
| | $x_4$ | 6 | 1 | 1 | | 1 | | | | |
| | $x_7$ | 2 | 1 | -1 | | | -1 | | 1 | |
| | $x_8$ | 3 | 1 | 1 | | | | -1 | | 1 |
| | $w$ | -5 | 2 | | | | -1 | -1 | | |
| 2 | $x_3$ | 7 | | 4 | 1 | | 1 | | -1 | |
| $s=1$ | $x_4$ | 4 | | 2 | | 1 | 1 | | -1 | |
| $r=7$ | $x_1$ | 2 | 1 | -1 | | | -1 | | 1 | |
| | $x_8$ | 1 | | 2 | | | 1 | -1 | -1 | 1 |
| | $w$ | -1 | | 2 | | | 1 | -1 | -2 | |
| 3 | $x_3$ | 5 | | | 1 | | -1 | 2 | 1 | -2 |
| $s=2$ | $x_4$ | 3 | | | | 1 | 1 | 1 | | -1 |
| $r=8$ | $x_1$ | $\tfrac{5}{2}$ | 1 | | | | $-\tfrac{1}{2}$ | $-\tfrac{1}{2}$ | $\tfrac{1}{2}$ | $\tfrac{1}{2}$ |
| | $x_2$ | $\tfrac{1}{2}$ | | 1 | | | $\tfrac{1}{2}$ | $-\tfrac{1}{2}$ | $-\tfrac{1}{2}$ | $\tfrac{1}{2}$ |
| | $w$ | 0 | | | | | | | -1 | -1 |

**4** For simplicity, consider the 3-variable problem:

Maximize $c_1x_1 + c_2x_2 + c_3x_3$

subject to $a_{11}x_1 + a_{12}x_2 + a_{13}x_3 = b_1$

$$a_{21}x_1 + a_{22}x_2 + a_{23}x_3 = b_2$$

$$x_1, x_2, x_3 \geqq 0.$$

The calculus and Lagrange multipliers cannot be used to solve this problem because of the inequality constraints. However, if we introduce new quantities $z_1$, $z_2$, $z_3$, and the additional constraints

$$x_1 = z_1^2$$

$$x_2 = z_2^2$$

$$x_3 = z_3^2$$

we can assume $x_1$, $x_2$, $x_3$ are positive. If we now introduce 5 Lagrange multipliers $y_i$, $i = 1, ..., 5$ the combined objective function becomes

$$H(X, Z, Y) = c_1x_1 + c_2x_2 + c_3x_3 + y_1(b_1 - a_{11}x_1 - a_{12}x_2 - a_{13}x_3)$$
$$+ y_2(b_2 - a_{21}x_1 - a_{22}x_2 - a_{23}x_3)$$
$$+ y_3(x_1 - z_1^2)$$
$$+ y_4(x_2 - z_2^2)$$
$$+ y_5(x_3 - z_3^2)$$

Differentiating,

$$\frac{\partial H}{\partial x_1} = c_1 - y_1a_{11} - y_2a_{21} + y_3 \qquad = 0$$

$$\frac{\partial H}{\partial x_2} = c_2 - y_2a_{12} - y_2a_{22} \qquad + y_4 \qquad = 0$$

$$\frac{\partial H}{\partial x_3} = c_3 - y_1a_{13} - y_2a_{23} \qquad\qquad + y_5 = 0$$

and the remaining derivatives produce the constraint equations with the additional restrictions that

$$y_3z_1 = 0$$
$$y_4z_2 = 0$$
$$y_5z_3 = 0.$$

However, we now face a difficulty: the $x_i$ quantities are absent from all these equations. All we have done is to create another problem in new variables $y_i$. Therefore the Lagrange multiplier method does not solve the problem. However, if we could solve the system of equations in $y_i$ to obtain solutions $y_i^*$ the value of $H(X, Z, Y)$ at the optimum will be

$$y_1^*b_1 + y_2^*b_2$$

as the coefficient in $x_1$ and $x_2$ and $x_3$ of $H(X, Z, Y)$ will be zero and $y_3z_1 = y_4z_2 = y_5z_3 = 0$.

That means at the optimum say $(x_1^*, x_2^*, x_3^*)$.

$$y_1^*b_1 + y_2^*b_2 = c_1x_1^* + c_2x_2^* + c_3x_3^*.$$

This is an interesting equivalence. It may also be noted that the system of

equations $\dfrac{\partial H}{\partial x_i}$ are remarkably like the standard linear programming problem

with slack variables added. In fact all these ideas form the basis of duality in linear programming which will be discussed later in Chapter 11.

## Chapter 6 (p. 98)

**1** $\dfrac{\partial F}{\partial x_1} = 3x_1^2 + 6x_1 - 2x_2^2$

$\dfrac{\partial F}{\partial x_2} = -4x_1 x_2.$

At $(1, 2)$, $\dfrac{\partial F}{\partial x_1} = 1$, $\dfrac{\partial F}{\partial x_2} = -8$ giving the steepest descent vector as

$\left( -\dfrac{1}{\sqrt{65}}, \dfrac{8}{\sqrt{65}} \right).$

At $(1, 1)$, $\dfrac{\partial F}{\partial x_1} = 7$, $\dfrac{\partial F}{\partial x_2} = -4$ giving the steepest descent vector as

$\left( -\dfrac{7}{\sqrt{65}}, \dfrac{4}{\sqrt{65}} \right).$

At $(2, 1)$, $\dfrac{\partial F}{\partial x_1} = 22$, $\dfrac{\partial F}{\partial x_2} = -8$ giving the steepest descent vector as

$\left( -\dfrac{11}{\sqrt{137}}, \dfrac{4}{\sqrt{137}} \right).$

**2** At $x^{(0)} = 0$, $-\dfrac{\partial F}{\partial x} = +7$, which as a unit vector is 1. Take $l = 1$ initially,

then

$F(0) = 9; F(1) = 3 < 9; F(2) = -1 < 3; F(4) = -3 < -1; F(8) = 17 > -3.$

Next point $x^{(1)} = 4$, $-\dfrac{\partial F}{\partial x}$ at $x = 4$ is $-1$

$F(4) = -3; F(3) = -3.$

As there is no reduction, halve step length $l/2$;

$F(3\tfrac{1}{2}) = -3 \cdot 25 < -3.$

Therefore $x^{(2)} = 3\tfrac{1}{2}$ and this is the minimum to an accuracy of $0 \cdot 5$.

**3** The partial derivatives are

$$\frac{\partial F}{\partial x_1} = 2x_1 + x_2 + 3x_1^2 + 2x_1 x_2$$

$$\frac{\partial F}{\partial x_2} = x_1 + 2x_2 + x_1^2 + 3x_2^2$$

and the direction of steepest descent from the point (1, 1) is

$$\left( \frac{-8}{\sqrt{113}}, \frac{-7}{\sqrt{113}} \right).$$

If $l$ denotes the step length, we require to reach the origin in one iteration

$$\left( 1 - \frac{8l}{\sqrt{113}}, 1 - \frac{7l}{\sqrt{113}} \right) = (0, 0)$$

i.e.

$$l = \frac{\sqrt{113}}{8} \text{ and } l = \frac{\sqrt{113}}{7}.$$

Since $l$ cannot equal both values, it is impossible to reach the origin in one step.

**4** Start from the point $X^{(0)} = (0, 4)$. The gradient there is

$$\left( -\frac{\partial F}{\partial x_1}, -\frac{\partial F}{\partial x_2} \right) = (10, 0).$$

The direction of steepest descent $D^{(1)} = (1, 0)$. The distance we can proceed is

$$L^{(0)} = \min \left[ \frac{10-4}{2}, \frac{4}{1} \right] = 3.$$

Therefore $X^{(1)} = (0, 4) + 3(1, 0) = (3, 4)$, at which point the first constraint is binding.

*Second iteration*

For the second iteration $\left( -\frac{\partial F}{\partial x_1}, -\frac{\partial F}{\partial x_2} \right) = (4, 0)$ and $D^{(2)} = (1, 0)$

$$P^{(2)} = \begin{pmatrix} 1 & 0 \\ 0 & 1 \end{pmatrix} - \begin{pmatrix} 2 \\ 1 \end{pmatrix} \left[ (2, 1) \begin{pmatrix} 2 \\ 1 \end{pmatrix} \right]^{-1} (2, 1) = \begin{bmatrix} \frac{1}{5} & -\frac{2}{5} \\ -\frac{2}{5} & \frac{4}{5} \end{bmatrix}.$$

Hence $P^{(2)} \cdot D^{(2)} = \left( \frac{1}{5}, \frac{-2}{5} \right)$ which is $\left( \frac{1}{\sqrt{5}}, \frac{-2}{\sqrt{5}} \right)$ as a unit vector. Therefore

$$X^{(2)} = (3, 4) + L^{(2)} \left( \frac{1}{\sqrt{5}}, \frac{-2}{\sqrt{5}} \right), \text{ where } L^{(2)} \text{ is the distance we move along the}$$
direction $D^{(2)}$.

As $L^{(2)} = \sqrt{5}$, $X^{(2)} = (4, 2)$.

**5** *Third iteration*

We are at the point (4, 2). The direction of steepest descent is (2, 4) which

gives $D^{(3)} = \left(\dfrac{1}{\sqrt{5}}, \dfrac{2}{\sqrt{5}}\right)$ as a unit vector. Both constraints are binding here so

$$M = \begin{pmatrix} 2 & 1 \\ 1 & 0 \end{pmatrix}$$

and the projection matrix $P^{(3)}$ is

$$\begin{pmatrix} 1 & 0 \\ 0 & 1 \end{pmatrix} - \begin{pmatrix} 2 & 1 \\ 1 & 0 \end{pmatrix} \begin{bmatrix} 5 & 2 \\ 2 & 1 \end{bmatrix}^{-1} \begin{pmatrix} 2 & 1 \\ 1 & 0 \end{pmatrix} = \begin{pmatrix} 0 & 0 \\ 0 & 0 \end{pmatrix}.$$

Also $(M'M)^{-1}M' . D^{(3)} = (2/\sqrt{5}, -4/\sqrt{5})$, indicating that the second constraint can be dropped.

*Fourth iteration*

For the next iteration $D^{(4)} = D^{(3)}$ and

$$M = \begin{pmatrix} 2 \\ 1 \end{pmatrix}$$

giving

$$P^{(4)} = \begin{pmatrix} \frac{1}{5} & -\frac{2}{5} \\ -\frac{2}{5} & \frac{4}{5} \end{pmatrix}.$$

Hence $P^{(4)} . D^{(4)} = \dfrac{1}{5\sqrt{5}} (-3, 6)$ which is $(-1/\sqrt{5}, 2/\sqrt{5})$ as a unit vector.

This vector is pointing back along the first constraint in the opposite direction which indicates that the minimum has been overshot. The intermediate points

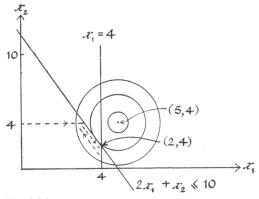

Fig. A6.1

along this line must therefore be inspected in the manner used for steepest descent. The diagram of the problem and the track followed is shown.

**6** $P(X, r) = x_1^2 + 2x_1x_2 + x_2^3 - \dfrac{r}{5x_1 + 3x_2 - 14} - \dfrac{r}{x_1 + x_2 - 7} + \dfrac{1}{\sqrt{r}}(x_1^2 + x_2^2 - 3)^2.$

**7** The response surface is

$$P(x, r) = x + \frac{r}{x} + \frac{1}{\sqrt{r}}(x - 2)^2.$$

For $r = 1$, $P(x, r) = x + \dfrac{1}{x} + (x - 2)^2.$

For $r = \frac{1}{16}$, $P(x, r) = x + \dfrac{1}{16x} + 4(x - 2)^2.$

The form of the curves are illustrated in Fig. A6.2. (They can be readily obtained by summing the curves of the separate components.)

The function $P(x, r)$ is differentiated to give the equation

$$\frac{dP}{dx} = 1 - \frac{r}{x^2} + \frac{2}{\sqrt{r}}(x - 2) = 0.$$

This may be written as

$$x - 2 = \frac{\sqrt{r}}{2}\left(\frac{r}{x^2} - 1\right).$$

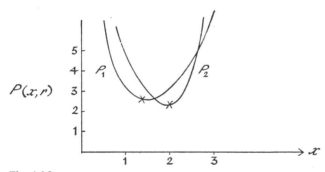

$P(x, r)$

Fig. A6.2

When $r$ becomes a very small number, the right-hand side of the equation will become a very small negative number implying that $x$ is very close to the value 2 and just less than 2.

**8** Evaluate base $B^{(0)} = (0, 0)$, $F(0, 0) = 50$.

Explore around base $B^{(0)} = T_0^{(0)}$, $F(1, 0) = 45$     accept

$F(1, 1) = 41$     accept

New base $B^{(1)} = (1, 1)$

Temporary position $T_0^{(1)} = (2, 2)$ with $F(2, 2) = 34$

---

Explore around $T_0^{(1)}$,  $\quad\quad\quad$ $F(3, 2) = 29$  $\quad$ accept

$\quad\quad\quad\quad\quad\quad\quad\quad\quad\quad$ $F(3, 3) = 29$  $\quad$ reject

$\quad\quad\quad\quad\quad\quad\quad\quad\quad\quad$ $F(3, 1) = 35$  $\quad$ reject

New base $B^{(2)} = (3, 2)$

Temporary position $T_0^{(2)} = (6, 4) - (1, 1) = (5, 3)$ with $F(5, 3) = 23$

---

Explore around $T_0^{(2)}$,  $\quad\quad\quad$ $F(6, 3) = 26$  $\quad$ reject

$\quad\quad\quad\quad\quad\quad\quad\quad\quad\quad$ $F(4, 3) = 24$  $\quad$ reject

$\quad\quad\quad\quad\quad\quad\quad\quad\quad\quad$ $F(5, 4) = 21$  $\quad$ accept

New base $B^{(3)} = (5, 4)$

Temporary position $T_0^{(3)} = (10, 8) - (3, 2) = (7, 5)$ with $F(7, 5) = 19$

---

Explore around $T_0^{(3)}$,  $\quad\quad\quad$ $F(8, 5) = 22$  $\quad$ reject

$\quad\quad\quad\quad\quad\quad\quad\quad\quad\quad$ $F(6, 5) = 20$  $\quad$ reject

$\quad\quad\quad\quad\quad\quad\quad\quad\quad\quad$ $F(7, 6) = 21$  $\quad$ reject

$\quad\quad\quad\quad\quad\quad\quad\quad\quad\quad$ $F(7, 4) = 23$  $\quad$ reject

New base $B^{(4)} = (7, 5)$

Temporary position $T_0^{(4)} = (14, 10) - (5, 4) = (9, 6)$ with $F(9, 6) = 23$

---

Explore around $T_0^{(4)}$,  $\quad\quad\quad$ $F(10, 6) = 30$  $\quad$ reject

$\quad\quad\quad\quad\quad\quad\quad\quad\quad\quad$ $F(8, 6) = 20$  $\quad$ accept

$\quad\quad\quad\quad\quad\quad\quad\quad\quad\quad$ $F(8, 7) = 24$  $\quad$ reject

$\quad\quad\quad\quad\quad\quad\quad\quad\quad\quad$ $F(8, 5) = 22$  $\quad$ reject

New base $B^{(5)} = B^{(4)} = (7, 5)$

Temporary position $T^{(5)} = (7, 5)$, with $F(7, 5) = 19$

---

As a tentative search has already been constructed around $(7, 5)$ it is known that this is the optimum.

The exact optimum can be found by solving the derivative equations:

$$\frac{\partial F}{\partial x_1} = 4x_1 - 7 - 4x_2 = 0$$

$$\frac{\partial F}{\partial x_2} = 6x_2 - 3 - 4x_1 = 0$$

giving $x_1 = 6{\cdot}75$ and $x_2 = 5$ as the exact minimum.

This could be reached by the direct search method if the step length was now reduced to $\frac{1}{4}$.

**9** The two functions are $x_1^3 - 6x_1^2 + 9x_1$ and $2x_2 + 3$. The function

$$x_1^3 - 6x_1^2 + 9x_1$$

has the form shown in the diagram and the four segments are illustrated, the dividing points being 0, 1, 3, 4 and 5.

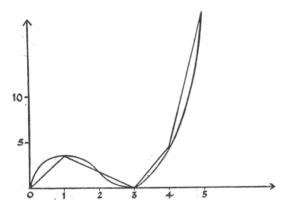

Fig. A6.3

The function is expressed in linear form as

$$\hat{f}_1(x_1) = 0.l_{11} + 4l_{12} + 0.l_{13} + 4.l_{14} + 20l_{15}.$$

The function $f_2(x_2)$ is simply a straight line and can be expressed exactly over the interval (1, 3) as

$$\hat{f}_2(x_2) = 5l_{21} + 9l_{22}.$$

The total function is expressed linearly in the $l$'s as:

$$\hat{F}(x_1, x_2) = 4l_{12} + 4l_{14} + 20l_{15} + 5l_{21} + 9l_{22}$$

and the $l$'s are constrained such that

$$l_{11} + l_{12} + l_{13} + l_{14} + l_{15} = 1$$
$$l_{21} + l_{22} = 1$$
$$l_{ij} \geq 0 \text{ for all } i \text{ and } j,$$

and for given $i$, if there are two $j$ values for which $l_{ij} > 0$ they must be adjacent.

**10** Starting from the point (3, 3), the approximating problem is:

Minimize $2x_1 + x_2$

subject to $6x_1 + 6x_2 \leqq 37$

$\qquad 6x_1 - 6x_2 \leqq 7$

$\qquad -1 \leqq x_1 - 3 \leqq 1$

$\qquad -1 \leqq x_2 - 3 \leqq 1.$

The solution is (4, 19/6).

If two further iterations were carried out, it would be found that the solutions were (3·993, 3·025) and (4·0, 3·0) which is the optimum.

**11** Clearly if the proposed step length $l$ is a small number compared with the step length actually taken it is probably worth while to set off with a longer trial step at the next iteration. Equally if $l$ is comparatively large it may be reduced. Let $l^{(t)}$ denote the trial step length for iteration $t$ and $L^{(t)}$ denote the actual step length taken at iteration $t$. Then a simple scheme for dynamically adapting $l$ as the iterations proceed would be to determine the step length $l^{(t)}$ at iteration $t$ as half the actual step length taken at the previous iteration, i.e. $l^{(t)} = \frac{1}{2}L^{(t-1)}$.

**12** The Taylor series expansion to three terms can be used to express the value of the function $F(X^{(0)})$ at the point $(X^{(0)} + l . D^{(0)})$ where $D^{(0)}$ is the steepest descent vector at $X^{(0)}$, as

$$F(X^{(0)} + lD^{(0)}) = F(X^{(0)}) + l \sum_{i=1}^{N} \frac{\partial F}{\partial x_i} . d_i^{(0)} + \frac{1}{2} l^2 \sum_{i=1}^{N} \sum_{j=1}^{N} \frac{\partial^2 F}{\partial x_i \partial x_j} d_i^{(0)} d_j^{(0)}$$

where all derivatives are evaluated at $X = X^{(0)}$ and $D^{(0)} = (d_1^{(0)}, d_2^{(0)}, ..., d_N^{(0)})$. We wish to determine $l$ so as to minimize $F(X^{(0)} + lD^{(0)})$. Therefore differentiating the right-hand side of the equation with respect to $l$ and equating to zero gives

$$l = \frac{- \sum_{i=1}^{N} (d_i^{(0)})^2}{\sum_{i=1}^{N} \sum_{j=1}^{N} \frac{\partial^2 F}{\partial x_i \partial x_j} . d_i^{(0)} . d_j^{(0)}}.$$

Although this is an automatic way of determining $l$ which eliminates the need for a search, the procedure will not always converge to the optimum. This difficulty is discussed in texts on numerical analysis.

If this formula is applied to Example 6.1 we obtain

$$l = \frac{-(4^2 + 24^2)}{2(4)^2 + 8(24)^2}.$$

As $(d_1^{(0)}, d_2^{(0)}) = (-4, -24)$, the new point we move to as $l \cdot D^{(0)}$, is approximately $(0.5, 3.0)$ which is a step length of just over 3 agreeing with the steepest descent results.

**13** A 2-variable linear programming problem is illustrated in Fig. A6.4.

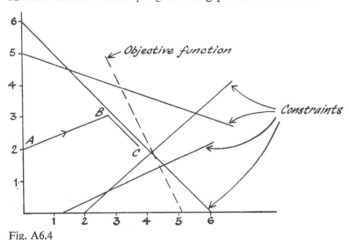

Fig. A6.4

Starting from the point $A$ the projected gradient method would proceed straight across the feasible region to the point $B$. It would then be necessary to move down the constraint to the optimum $C$. Although this may seem to be an efficient way of reaching the optimum it is not generally better than the simplex method which works round the edge of the region. The amount of calculation at each iteration is greater, and unless the starting point in the simplex method happens to be on the other side of the feasible region—in which case the projected gradient would go straight across the region—both methods will consist of working their way around the region from vertex to vertex. The initial point will not generally be at the vertex diametrically opposite the optimum so that the simplex method is usually preferable.

**14** A system of contours is illustrated in which the track which would be followed by the projected gradient method would be $A$, $B$, $C$, $D$ as illustrated in Fig. A6.5. $A$ is the initial point and the steepest descent method moves to position $B$. We then move along the constraint to $C$ and finally turn in to the feasible region to the minimum at $D$.

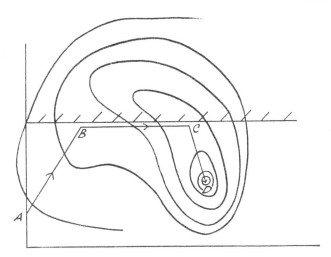

Fig. A6.5

**15** The points $x_i^{(j)}$ could be determined for the function $f_i(x_i)$ as the local maxima, minima and points of inflexion of the curve $f_i(x_i)$ in the range $(b_i, B_i)$. They would be determined as the roots of the equation

$$\frac{df_i}{dx_i} = 0.$$

## Chapter 7 (p. 120)

**1** Defining stages 1, 2, 3, 4, 5, 6 as containing the points (1), (2, 3), (4, 5), (6, 7), (8, 9), (10), respectively, the recurrence relations of Section 7.5 can be used to obtain the following results. Given $f_1(1) = 0$

| | |
|---|---|
| $f_2(2) = 1$ | $q_1(2) = 1$ |
| $f_2(3) = 2$ | $q_1(3) = 1$ |
| $f_3(4) = \min(1+6, 2+3) = 5$ | $q_2(4) = 3$ |
| $f_3(5) = \min(6, 7) = 6$ | $q_2(5) = 2$ |
| $f_4(6) = \min(9, 12) = 9$ | $q_3(6) = 4$ |
| $f_4(7) = \min(10, 9) = 9$ | $q_3(7) = 5$ |
| $f_5(8) = \min(11, 10) = 10$ | $q_4(8) = 7$ |
| $f_5(9) = \min(12, 11) = 11$ | $q_4(9) = 7$ |
| $f_6(8) = \min(15, 14) = 14$ | $q_5(10) = 9$ |

Hence the shortest distance is 14 and the route is 1, 2, 5, 7, 9, 10.

**2** Define $f_k(y)$ as the optimum allocation if a total of $y$ units are used on the first $k$ operations, and $q_k(y)$ as the allocation to operation $k$. Then

$$f_k(y) = \max_{0 \le x_k \le y} (g_k(x_k) + f_{k-1}(y - x_k)).$$

This gives the following table:

**Table A7.1**

| $y$ | $f_1(y)$ | $q_1(y)$ | $f_2(y)$ | $q_2(y)$ | $f_3(y)$ | $q_3(y)$ | $f_4(y)$ | $q_4(y)$ |
|---|---|---|---|---|---|---|---|---|
| 0 | 2 | 0 | 2 | 0 | 3 | 0 | 3 | 0 |
| 1 | 4 | 1 | 4 | 0 | 5 | 0 | 6 | 1 |
| 2 | 6 | 2 | 6 | 0 | 7 | 1 | 9 | 2 |
| 3 | 7 | 3 | 7 | 0 | 9 | 1 | 11 | 2 |
| 4 | 7 | 4 | 8 | 1 | 11 | 2 | 13 | 2 |
| 5 | 7 | 5 | 9 | 2 | 13 | 3 | 15 | 2 |
| 6 | 7 | 6 | 10 | 3 | 14 | 3 | 17 | 2 |

Therefore the maximum return is 17, and it is obtained by allocating 2 units to operations 1 and 3 and 4.

If the supply was reduced to 3 units, the maximum return would be 11 obtained by allocating 2 units to operation 4, and 1 unit to operation 1.

**3** Using the recurrence relations of Section 7.6 we obtain the following table:

**Table A7.2**

| $y$ | STAGE 1 $f_1(y)$ | $q_1(y)$ | STAGE 2 $f_2(y)$ | $q_2(y)$ | STAGE 3 $f_3(y)$ | $q_3(y)$ | STAGE 4 $f_4(y)$ | $q_4(y)$ |
|---|---|---|---|---|---|---|---|---|
| 0 | 0 | 0 | 0 | 0 | 0 | 0 | 0 | 0 |
| 1 | 1 | 1 | 1 | 0 | 1 | 0 | 1 | 0 |
| 2 | 2 | 2 | 2 | 0 | 2 | 0 | 2 | 0 |
| 3 | 3 | 3 | 5 | 1 | 5 | 0 | 5 | 0 |
| 4 | 4 | 4 | 6 | 1 | 7 | 1 | 7 | 0 |
| 5 | 5 | 5 | 7 | 1 | 8 | 1 | 8 | 0 |
| 6 | 6 | 6 | 10 | 2 | 10 | 0 | 11 | 1 |
| 7 | 7 | 7 | 11 | 2 | 12 | 1 | 12 | 0 |
| 8 | 8 | 8 | 12 | 2 | 14 | 2 | 14 | 0 |
| 9 | 9 | 9 | 15 | 3 | 15 | 2 | 16 | 1 |
| 10 | 10 | 10 | 16 | 3 | 17 | 1 | 18 | 1 |
| 11 | 11 | 11 | 17 | 3 | 19 | 2 | 19 | 0 |

This shows that the maximum value is 19 and the cargo contains 2 items of type 3 and 1 item of type 2.

**4** Using the recurrence relations of Section 7.7, we determine the $f_k(y)$ values as follows:

**Table A7.3**

| $y$ | STAGE 1 $f_1(y)$ | $q_1(y)$ | STAGE 2 $f_2(y)$ | $q_2(y)$ | STAGE 3 $f_3(y)$ | $q_3(y)$ |
|---|---|---|---|---|---|---|
| 1 | $\frac{1}{2}$ | 1 | 0 | 0 | 0 | 0 |
| 2 | $\frac{3}{4}$ | 2 | 0 | 0 | 0 | 0 |
| 3 | $\frac{7}{8}$ | 3 | $\frac{1}{4}$ | 1 | 0 | 0 |
| 4 | $\frac{15}{16}$ | 4 | $\frac{3}{8}$ | 1 | 0 | 0 |
| 5 | $\frac{31}{32}$ | 5 | $\frac{7}{16}$ | 1 | 0 | 0 |
| 6 | $\frac{63}{64}$ | 6 | $\frac{9}{16}$ | 2 | $\frac{1}{8}$ | 1 |
| 7 | $\frac{127}{128}$ | 7 | $\frac{21}{32}$ | 2 | $\frac{3}{16}$ | 1 |
| 8 | $\frac{255}{256}$ | 8 | $\frac{45}{64}$ | 2 | $\frac{7}{32}$ | 1 |
| 9 | $\frac{511}{512}$ | 9 | $\frac{49}{64}$ | 3 | $\frac{9}{32}$ | 1 |

The maximum reliability of the system is $\frac{9}{32}$ when 1 component of type 3 and 2 components of types 1 and 2 are included.

**5** Determine values for $x_1, x_2, ..., x_N$ as an $N$-stage process and define $f_k(y)$ as the maximum value of $\sum_{i=1}^{k} g_i(x_i)$ where $x_1, x_2, x_3, ..., x_k = y$.

Then the recurrence relations are:

$$f_k(y) = \max_{1 \leq x_k \leq y} \left[ g_k(x_k) + f_{k-1}\left(\frac{y}{x_k}\right) \right]$$

and $f_1(y) = y$. These are evaluated for $k = 1, 2, ..., N$.

**6** Define $f_k(y)$ as the minimum of

$$\max (g_1(x_1), g_2(x_2), ..., g_k(x_k))$$

where $x_1 + x_2 + ... + x_k = y$.

Then the recurrence relations are

$$f_k(y) = \min_{0 \leq x_k \leq y} [\max (g_k(x_k), f_{k-1}(y - x_k))]$$

where $f_1(y) = g_1(y)$.

Notice that the solution to this problem could not be provided by any of the optimization procedures so far discussed.

**7** Define $f_k(y)$ as $\min (x_1^2 + x_2^2 + ... + x_k^2)$, subject to $x_1 + x_2 + ... + x_k = y$.

Then

$$f_1(y) = y^2$$

$$f_2(y) = \min_{0 \leq x_2 \leq y} (x_2^2 + f_1(y - x_2))$$

$$= \min_{0 \leq x_2 \leq y} (x_2^2 + (y - x_2)^2).$$

Differentiating the function $x_2^2 + (y - x_2)^2$ with respect to $x_2$ gives a minimum at

$x_2 = y/2.$

Therefore

$f_2(y) = y^2/2.$

$$f_3(y) = \min_{0 \leq x_3 \leq y} (x_3^2 + f_2(y - x_3))$$

$$= \min_{0 \leq x_3 \leq y} \left( x_3^2 + \frac{(y - x_3)^2}{2} \right).$$

Differentiation gives

$x_3 = y/3.$

Hence the optimum occurs for $x_1 = x_2 = x_3 = C/3.$

**8** Starting from point 1 we can regard all points which can be reached in one step as the states of stage 1, all points which can be reached in **at most two** steps as the states of stage 2 and so on. (There is no need actually to construct this new network—it merely needs to be visualized.) Let $f_k(j)$ denote the shortest route from point 1 to point $j$ in at most $k$ steps. Then initially we set $f_0(1) = 0$ and otherwise $f_0(j) = \infty$. Then the recurrence relations are

$$f_k(j) = \min_{1 \leq i \leq N} \{d(i, j) + f_{k-1}(i)\}$$

and this must be evaluated for $j = 1, ..., N$ and $k = 1, ..., n$, where $n$ is the first iteration on which all $f_k(j)$ remain unchanged. We can record as $q_{k-1}(j)$ the value of the point for which the minimization holds.

**9** The stages of the travelling salesman problem may be regarded as the series of cities which have been visited. We regard the states at each stage as the various possible collections of cities which may have been visited up to this stage. Define $f_k(x_1, x_2, ..., x_k; i)$ as the optimal tour for the remaining $k$ cities having reached city $i$ and after visiting all cities other than $x_1, x_2, ..., x_k$. Then if $d(i, j)$ is the distance from city $i$ to city $j$ the recurrence relations are

$$f_k(x_1, x_2, ..., x_k; i) = \min_{1 \leq r \leq k} \{d(i, x_r) + f_{k-1}(x_1, x_2, ..., x_{r-1}, x_{r+1}, ..., x_k; x_r)\}$$

where $x_1, x_2, ..., x_k$ is a selection from the total of $N$ cities. Unfortunately the number of possible selections which must be recorded at stage $k$ is

$\dfrac{N!}{(N-k)!k!}$ which will be a very large number. However, the dynamic programming formulations has reduced the problem into combinations rather than permutations.

**10** The recurrence relation for the resource allocation problem of Section 7.6 was

$$f_k(y) = \max_{0 \le x_k \le y} \{g_k(x_k) + f_{k-1}(y - x_k)\}.$$

If we drop the assumption that the $g_k(x_k)$ functions are monotonic it is no longer necessarily true that the $f_k(y)$ functions will be monotonic. This leads to the difficulty that if $x_k$ is allocated to the $k$th activity it is no longer true that the best allocation to the first $(k-1)$ activities is $(y - x_k)$. The recurrence equation therefore requires a further dimension of maximization and this makes the whole computation more complex.

**11** Let us start from the end in this problem without knowing how many stages we have to work through. The states at each stage correspond to the possible levels in the containers and the action of pouring contents between containers or filling up from the bag corresponds to moving from stage to stage.

At the last stage there are 7 grams in the 8-gram container because there is no other possible solution.

At the previous stage, one possible state which could lead to the desired final state is having 2 grams in the 8-gram container and 5 grams in the 5-gram container, as then the final state is reached by pouring the contents of the 5-gram container into the 8-gram container.

At the next stage back we require to be able to move into the state in which the 5-gram container has 5 grams and the 8-gram container has 2. The 5-gram container should therefore be empty and we should move along the track which leads to it being completely filled up from the bag at the next stage. Let us go back a further stage to see how the 8-gram container can have 2 grams.

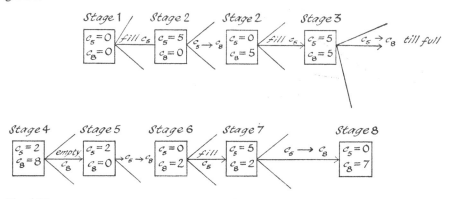

Fig. A7.1

Clearly this must be done by pouring some material out of a container. Suppose the 5-gram container is the one which holds the 2 grams and this is poured into the 8-gram container. How did the 5-gram container have 2 grams?

Presumably, by pouring 3 grams out of the 5-gram container at the previous stage.

But how would we know when to stop? Only if there were already 5 grams in the 8-gram container.

This could have been done at the previous stage by pouring a full 5 grams from the 5-gram container into the 8-gram container.

Thus the decision sequence can be represented in stages and states as follows. We denote by $c_5$, $c_8$ the contents of the 5- and 8-gram containers respectively and record the actions linking the states on the arrows connecting the states.

This is not exactly an optimization problem unless we wish to put a high return on obtaining $c_8 = 7$, $c_5 = 0$ at the final stage and a zero return otherwise. But it is a further illustration of the process of the multi-stage treatment of a problem which is the basic dynamic programming method.

## Chapter 8 (p. 139)

**1** For the two-machine scheduling problem we can use the following bounds $LB(n)$ similar to the three-machine bounds of Section 8.3,

$$LB(n) = \max \begin{cases} T_a(S_n) + \sum_j a_j + \min_j b_j \\ T_b(S_n) + \sum_j b_j \end{cases} \quad j \in E_n$$

where $S_n$ are the scheduled jobs at node $n$ and $E_n$ are the excluded jobs. If this method is applied to the example data, we obtain the following tree solution:

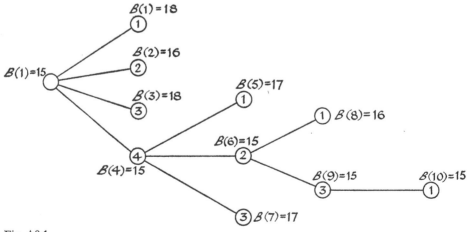

Fig. A8.1

Therefore the optimum sequence is 4, 2, 3, 1 and the production time is 15 as shown in Fig. A8.2.

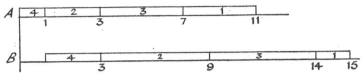

Fig. A8.2

**2** Applying the knapsack procedure of Section 8.4 to the data we obtain the following tree:

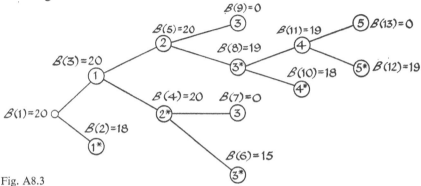

Fig. A8.3

The optimum collection of items is indicated by the branch leading to node 12; it contains items 1, 2 and 4 and has a value of 19.

**3** The constraints limit $x_1$ to lie below 3 and $x_2$ to lie below 5 and the optimum non-integer solution is at $x_1 = 2\frac{1}{2}$, $x_2 = 4\frac{1}{2}$. Using the second branching policy which was described in Example 8.3 the solution is enumerated by the tree of nodes shown. The optimum occurs at the node 2 with $x_1 = 3$, $x_2 = 2$.

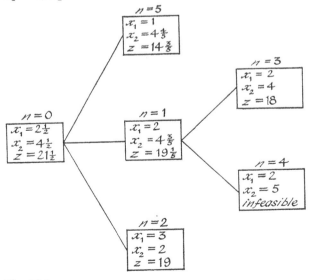

Fig. A8.4

**4** One method for solving the travelling salesman problem would be to proceed along the lines which have been described for the three-machine scheduling problem. Starting with city 1 in node 1 we try cities 2, 3, ..., $N$ in position 2. We then select one of these, say 4, with the least bound and try cities 2, 3, 5, ..., $N$ in position 3. We thus get a tree of the form shown in Fig. A8.5.

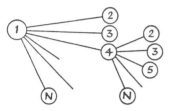

Fig. A8.5

Lower bounds could be estimated on a branch with $k$ points included, as the distance along the branch plus the $(N-k)$ smallest numbers in the reduced distance matrix which excludes the rows and columns of the points within the branch, e.g. point 4 is in the branch 1, 4, 5 so that the row and column $d(4, j)$ and $d(i, 4)$ can be eliminated.

An alternative approach, developed by Little, operates like the procedure for the knapsack problem. We consider whether or not to include certain links in the tour. A link $(i, j)$ indicates that city $j$ is to be reached from city $i$. Starting with all possible solutions, we select the link say $(i, j)$ which, if not included, will offer the largest lower bound. This subdivides the problem into two as shown, which include $(i, j)$ or exclude $(i, j)$ marked as $(i, j)$ and $(i, j^*)$.

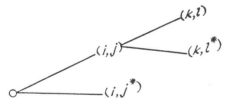

Fig. A8.6

We then go on to choose another city pair $(k, l)$ and consider its inclusion or exclusion, and continue in this way calculating lower bounds and advancing from the least lower bound. The next city pair $(i, j)$ for the tree is chosen to attempt to maximize the lower bound on $(i, j^*)$. This means that we are concentrating on eliminating potentially bad tours rather than trying to select good tours in this branch and bound policy. If the link $(r, s)$ is not included then we know that we must reach $s$ from some city other than $r$ and will go to some city other than $s$ from city $r$. Thus we can add up the two least distances leading out of $r$ and leading into $s$ excluding $(r, s)$ in both cases and this gives us a measure of cost say $C(r, s)$ of not including the link $(r, s)$. We choose $(i, j)$ as the next node which maximizes this direct cost.

The determination of the lower bounds on a node is also intricate. If the node $(i, j)$ has bound $B(i, j)$ then the lower bound corresponding to $(k, l^*)$ which is reached directly from $(i, j)$ is $B(i, j) + C(k, l)$ where $C(k, l)$ is calculated as described above. The bound on the node $(k, l)$ is an estimate of the minimum distance which must be added to the bound $B(i, j)$ when the rows and columns corresponding to the fixed included links and $(k, l)$ are removed. An estimate of this minimum distance is to find the smallest element in each row of the reduced distance matrix and subtract it from all elements of the row, and then do the same for columns. The sum of all the subtracted elements is a valid and fairly good lower bound on the optimal tour.

**5** As the jobs do not all go through the machines in the same sequence it is necessary to define a correspondence between nodes and operations rather than nodes and jobs. Thus if there are $N$ jobs we can define the first set of nodes as corresponding to the first operations of all the jobs. Each node extends to new nodes corresponding to the next operations of all jobs. If $(j, k)$ denotes the $k$th operation of job $j$ and there are three jobs, a tree development could take the form:

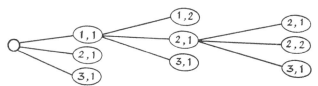

Fig. A8.7

The second column of nodes shows that the first operations of jobs 2 and 3 have still to be done, whereas the first operation of job 1 has been completed. A branch such as $(1, 1)$, $(2, 1)$, $(2, 2)$ would correspond to loading these operations on to the appropriate machines in this order. Bounds on node $n$ could be estimated by summing the total work remaining on each machine say $a(n)$, $b(n)$, $c(n)$ and adding these on to the earliest free time $A(n)$, $B(n)$, $C(n)$. The lower bound $B(n)$ corresponding to node $n$ would be

$$B(n) = \max\,(A(n) + a(n),\; B(n) + b(n),\; C(n) + c(n)).$$

This problem has less structure than the case where all jobs go through the same sequence and the bound formula is probably less efficient.

## Chapter 9 (p. 151)

**1** One case of a locally optimal circuit through these points is shown as the circuit $[1, 2, 5, 6, 3, 4]$ with length $5 + \sqrt{5}$.

A typical circuit one adjacent exchange away from this circuit is the permutation [1, 2, 5, 3, 6, 4] which has length $6+\sqrt{2}>5+\sqrt{5}$. The minimum circuit [1, 2, 3, 6, 5, 4] has length 6 units and cannot be reached by a single adjacent exchange.

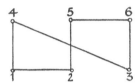

Fig. A9.1

**2** The initial ordering $[P^{(0)}]$ is [4, 1, 2, 3] with production time 18. The adjacent exchanges are now applied in turn to give the results

| Current permutation | Production time | Base |
|---|---|---|
| $[P^{(1)}] = [E_1 P^{(0)}] = [1, 4, 2, 3]$ | 18 | $[P^{(0)}]$ |
| $[P^{(2)}] = [E_2 P^{(0)}] = [4, 2, 1, 3]$ | 16 | $[P^{(2)}]$ |
| $[P^{(3)}] = [E_3 P^{(2)}] = [4, 2, 3, 1]$ | 15 | $[P^{(3)}]$ |
| $[P^{(4)}] = [E_1 P^{(3)}] = [2, 4, 3, 1]$ | 16 | $[P^{(3)}]$ |
| $[P^{(5)}] = [E_2 P^{(3)}] = [4, 3, 2, 1]$ | 17 | $[P^{(3)}]$ |
| $[P^{(6)}] = [E_3 P^{(3)}] = [4, 2, 1, 3]$ | 16 | $[P^{(3)}]$. |

The locally optimum permutation is therefore [4, 2, 3, 1]. In fact, this is also the global optimum as confirmed by the answers to Exercise 8.1.

**3** As the total loading time is constant and equal to four units we are simply concerned with minimizing the sum of the set-up times. There are four adjacent exchanges defined as

$E_1$ exchange elements $p_1$ and $p_2$

$E_2$ exchange elements $p_2$ and $p_3$

$E_3$ exchange elements $p_3$ and $p_4$

$E_4$ exchange elements $p_4$ and $p_5$.

Starting with $E_1$ applied to [1, 2, 3, 4, 5] we cycle through these exchanges until we have applied four exchanges without improvement.

| Exchange | Permutation | Set-up times | Base |
|---|---|---|---|
| | $P^{(0)} = [1, 2, 3, 4, 5]$ | $6+4+2+3 = 15$ | $P^{(0)}$ |
| $E_1$ | $P^{(1)} = [2, 1, 3, 4, 5]$ | $4+0+2+3 = 9$ | $P^{(1)}$ |

| | | | |
|---|---|---|---|
| $E_2$ | $P^{(2)} = [2, 3, 1, 4, 5]$ | $4+2+2+3 = 11$ | $P^{(1)}$ |
| $E_3$ | $P^{(3)} = [2, 1, 4, 3, 5]$ | $4+2+4+1 = 11$ | $P^{(1)}$ |
| $E_4$ | $P^{(4)} = [2, 1, 3, 5, 4]$ | $4+0+1+3 = 8$ | $P^{(4)}$ |
| $E_1$ | $P^{(5)} = [1, 2, 3, 5, 4]$ | $6+4+1+3 = 14$ | $P^{(4)}$ |
| $E_2$ | $P^{(6)} = [2, 3, 1, 5, 4]$ | $4+2+7+3 = 16$ | $P^{(4)}$ |
| $E_3$ | $P^{(7)} = [2, 1, 5, 3, 4]$ | $4+7+6+2 = 19$ | $P^{(4)}$ |
| $E_4$ | $P^{(8)} = [2, 1, 3, 4, 5]$ | $4+0+2+3 = 9$ | $P^{(4)}$. |

Hence the permutation $[2, 1, 3, 5, 4]$ is locally optimal and the total time is
28 units.

**4** The sequence on each machine must be represented by an individual permutation. Let $p_{ij}$ denote the $j$th job to be scheduled through the $i$th machine
and let $n(i)$ be the number of jobs on machine $i$. Then the permutation of jobs
on machine $i$ is denoted by

$$[P_i] = [p_{i1}, p_{i2}, ..., p_{in(i)}].$$

Let $d(i, j)$ denote the duration of job $j$ if it goes on the $i$th machine, let $t(j)$
be its due time and let $f(j, x)$ denote the cost of being $x$ units late. Then the
completion time of job $p_{ij}$ is

$$C(p_{ij}) = \sum_{k=1}^{j} d(i, p_{ik}),$$

and the total cost to be minimized is

$$F[P_1, P_2, P_3] = \sum_{i=1}^{3} \sum_{j=1}^{n(i)} f(p_{ij}, \max \{C(p_{ij}) - t(p_j), 0\}).$$

**5** Suppose the points are arranged as shown in the Fig. A9.2.

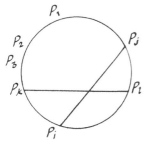

Fig. A9.2

Starting at any fixed point on the circle the order of the elements around the
circle may be denoted by

$$[p_1, p_2, ..., p_N].$$

H

Then point $p_i$ is connected to $p_j$ if $C(p_i, p_j) = 1$. Furthermore, if $C(p_k, p_l) = 1$ and $k$ lies on one side of the chord $(p_i, p_j)$ and $l$ lies on the other, there will be an intersection or crossing. Thus the total number of crossings on the chord $(p_i, p_j)$ is

$$\sum_{k=1}^{i-1} \sum_{l=i+1}^{j-1} C(p_k, p_l) + \sum_{k=j+1}^{N} \sum_{l=i+1}^{j-1} C(p_k, p_l).$$

Hence the total number of crossings to be minimized can be expressed in terms of the permutation $[P]$ as

$$F[P] = \tfrac{1}{2} \sum_{i=1}^{N-1} \sum_{j=i+1}^{N} C(p_i, p_j) \left\{ \sum_{k=1}^{i-1} \sum_{l=i+1}^{j-1} C(p_k, p_l) + \sum_{k=j+1}^{N} \sum_{l=i+1}^{j-1} C(p_k, p_l) \right\}.$$

The factor $\tfrac{1}{2}$ is necessary to eliminate the double counting. We will be using a formula similar to this in a later study on circuit design.

## Chapter 10 (p. 160)

**1** The difference $g(i)$ between the cheapest and second cheapest costs for each item $i$ are

$g_0(1) = 9-5 = 4$, $g_0(2) = 12-6 = 6$, $g_0(3) = 3-2 = 1$, $g_0(4) = 12-9 = 3$, $g_0(5) = 10-6 = 4$.

Select item 2 and allocate it to position 3 at a cost of 6.

$g_1(1) = 5$, $g_1(3) = 1$, $g_1(4) = 6$, $g_1(5) = 4$.

Select item 4 and allocate it to position 5 at a cost of 9.

$g_2(1) = 1$, $g_2(3) = 1$, $g_2(5) = 1$.

Select item 1 and allocate it to position 4 at a cost of 10.

$g_3(3) = 1$, $g_3(5) = 9$.

Select item 5 and allocate it to position 2 at a cost of 10. Finally allocate item 3 to position 1 at a cost of 4 units. This is the optimal allocation giving a total of 39.

**2** The items can be regarded as the units to be transported (although decisions can alternatively be made about the movement of groups of items). After $t$ decisions have been made we can find the cheapest route $(i, j)$ along which some supply and demand still remain to be transported. We can then allocate as much as possible along that route (i.e. the total remaining supply or demand, whichever is smaller).

The cheapest route is the route from supply point 4 to demand point 4, written as $(4, 4)$ with cost 2, and we assign 2 units along this route with total cost 4. The new cost matrix is:

**Table A10.1**

|   |   | 1<br>4 | 2<br>4 | 3<br>6 | 5<br>4 | 6<br>2 |
|---|---|---|---|---|---|---|
| 1 | 5 | 9 | 12 | 9 | 9 | 10 |
| 2 | 6 | 7 | 3 | 7 | 5 | 5 |
| 3 | 2 | 6 | 5 | 9 | 3 | 11 |
| 4 | 7 | 6 | 8 | 11 | 2 | 10 |

The cheapest route now is (4, 5) and 4 units are assigned with total cost 8.
   The new cost matrix is:

**Table A10.2**

|   |   | 1<br>4 | 2<br>4 | 3<br>6 | 6<br>2 |
|---|---|---|---|---|---|
| 1 | 5 | 9 | 12 | 9 | 10 |
| 2 | 6 | 7 | 3 | 7 | 5 |
| 3 | 2 | 6 | 5 | 9 | 11 |
| 4 | 3 | 6 | 8 | 11 | 10 |

The new cheapest route is (2, 2) with 4 units assigned at total cost 12.
   The new cost matrix is:

**Table A10.3**

|   |   | 1<br>4 | 3<br>6 | 6<br>2 |
|---|---|---|---|---|
| 1 | 5 | 9 | 9 | 10 |
| 2 | 2 | 7 | 7 | 5 |
| 3 | 2 | 6 | 9 | 11 |
| 4 | 3 | 6 | 11 | 10 |

The new cheapest route is (2, 6) and 2 units are assigned at a total cost of 10.
   The revised cost matrix is:
   H*

**Table A10.4**

|     | 1<br>4 | 3<br>6 |
|-----|--------|--------|
| 1 5 | 9      | 9      |
| 3 2 | 6      | 9      |
| 4 3 | 6      | 11     |

The new cheapest route is (3, 1) and 2 units are assigned at a total cost of 12. The revised cost matrix is:

**Table A10.5**

|     | 1<br>2 | 3<br>6 |
|-----|--------|--------|
| 1 5 | 9      | 9      |
| 4 3 | 6      | 11     |

The new cheapest route is (4, 1) and 2 units are assigned at a total cost of 12. The revised cost matrix is:

**Table A10.6**

|     | 3<br>6 |
|-----|--------|
| 1 5 | 9      |
| 4 1 | 11     |

5 units are allocated along route (1, 3) and 1 unit along route (4, 3), at a total cost of 56.

The collection of allocations are summarized in Table A10.7.

**Table A10.7**

|   | 1 | 2 | 3 | 4 | 5 | 6 |
|---|---|---|---|---|---|---|
| 1 |   |   | 5 |   |   |   |
| 2 |   | 4 |   |   |   | 2 |
| 3 | 2 |   |   |   |   |   |
| 4 | 2 |   | 1 | 2 | 4 |   |

The total cost is 114. The optimum cost is 112 and is obtained by re-allocating the first three columns as 0, 0, 1, 3; 0, 3, 1, 0; 5, 1, 0, 0.

**3** The nearest city approach would give the initial tour shown in Fig. A10.1.

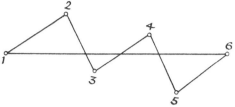

Fig. A10.1

The improved method of including the next best city in the sub-tour at the best position would build up the sub-tours as shown in Fig. A10.2.

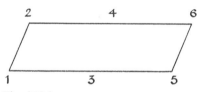

Fig. A10.2

giving a final optimal tour as shown in Fig. A10.3.

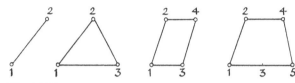

Fig. A10.3

**4** A convex boundary of points is illustrated in a 10-point travelling salesman problem in Fig. A10.4.

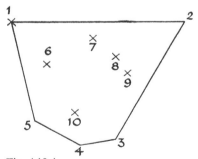

Fig. A10.4

To prove that the convex boundary will form a subsequence in any optimal circuit, first note that in an optimal circuit it is best to join all cities by straight lines and that no two lines will cross. For, if they do, it will always be possible to reduce the tour by uncrossing. An illustration is given in Fig. A10.5.

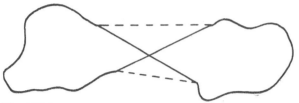

Fig. A10.5

Assuming there are $N$ points in total, suppose we number the points so that the convex boundary consists of the first $n$ of the $N$ points, i.e. the convex boundary is formed by the subsequence 1, 2, 3, ..., $n$. We will assume that all circuits are arranged so that city number 1 is in the first position. Suppose now that we have an optimal circuit which contains the points of the convex boundary in some order other than 1, 2, 3, ..., $n$. Then by checking this sequence, let $k(<n)$ be the last boundary point in this subsequence which is in the correct order, and let $j(\leq n)$ be the next boundary point visited after boundary point $k$. The situation is as illustrated in Fig. A10.6.

Now after reaching point $j$ it is necessary to continue the circuit to point $(k+1)$ and later return to point 1. Since the circuit is optimal, the path from point $(k+1)$ to 1 will consists of a series of straight lines, and this path will lie inside or on the convex boundary. However, the path from $k$ to $j$ cuts the convex boundary into two disjoint sections and points 1 and $(k+1)$ lie in

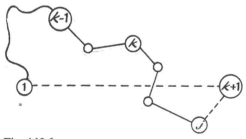

Fig. A10.6

different sections. Therefore the path from $(k+1)$ to 1 must cross the path from $k$ to $j$. This route therefore cannot be optimal since no crossings will occur on the optimal circuit. This contradicts the hypothesis, and therefore for the route to be optimal, the sequence of convex boundary points 1, 2, ..., $n$ must form a subsequence of the optimal tour.

The result can be used to obtain a good initial circuit for planar travelling salesmen problems by first finding the convex boundary and then steadily deforming it, to include one more city at a time until a complete circuit is obtained.

## Chapter 11 (p. 176)

**1** The two modified objective functions are

$$P(x) = x^2 - 4x + 4 + \frac{1}{10(x-1)},$$

$$Q(x) = x^2 - 4x + 4 + 10 \max(1-x, 0).$$

The curves for $F(x)$, $P(x)$ and $Q(x)$ are shown in Fig. A11.1. It should be noticed that $Q(x)$ has a much more stable behaviour than $P(x)$.

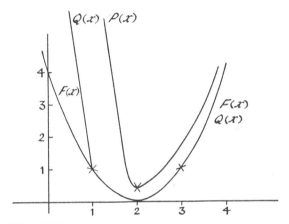

Fig. A11.1

**2** The problem can be rewritten as

Maximize $x_4$

subject to
$$2x_1 + 3x_2 + 6x_3 \leqq 12$$
$$4x_1 + 2x_2 + 5x_3 \leqq 15$$
$$-x_1 \qquad\qquad + x_4 \leqq 0$$
$$\qquad -x_2 \qquad + x_4 \leqq 0$$
$$\qquad\qquad -x_3 + x_4 \leqq 0.$$

Slack variables $x_5$, $x_6$, $x_7$, $x_8$, $x_9$, can now be introduced to convert the problem into a standard linear programming form.

**3** From the inequalities and the non-negativity constraints it is clear that

$$|x_1 - x_2| \leqq 20.$$

Introducing a zero-one variable $x_4$ we can convert the two conditional inequalities into 3 constraints as

$$(20+12)x_4+x_1-x_2 \geq 12$$
$$(20+8)(1-x_4)+x_2-x_1 \geq 8$$
$$0 \leq x_4 \leq 1, x_4 \text{ integral.}$$

The integer linear programming problem then becomes:

Maximize $5x_1+3x_2+x_3$

subject to
$$
\begin{array}{llll}
x_1-x_2 & +32x_4 & \geq 12 \\
-x_1+x_2 & -28x_4 & \geq -20 \\
x_1 & -x_3 & \leq 5 \\
& x_4 & \leq 1 \\
x_1, \ x_2, \ x_3, & x_4 \geq 0, & x_4 \text{ integral.}
\end{array}
$$

**4** Let $x_1 = y_1^2$

$x_2 = 5 \sin^2 y_2$

$x_3 = -4+10 \sin^2 y_3.$

Then the objective function in $y_1, y_2, y_3$ becomes

$F'(y_1, y_2, y_3) = F(x_1, x_2, x_3)$
$$= y_1^4+\sqrt{5} \sin y_2(y_1^2+10 \sin^2 y_3-4).$$

The constraints are all automatically satisfied as

$y_1^2 \geq 0$, for all values of $y_1$;

$0 \leq 5 \sin^2 y_2 \leq 5$ is equivalent to

$0 \leq \sin^2 y_2 \leq 1$ which is satisfied for all $y_2$;

$0 \leq 10 \sin^2 y_3 \leq 10$ holds for all $y_3$.

**5** The dual problem in the variables $y_1, y_2$ is

$2y_1+ y_2 \geq 8$

$3y_1+7y_2 \geq 21$

$y_1+ y_2 = v$ to be minimized

$y_1, y_2 \geq 0.$

By graphical solution it will be found that the solution to the primal is

$x_1 = \frac{4}{11}$

$x_2 = \frac{1}{11}$

$z = \frac{53}{11}$

and to the dual, the solution is

$$y_1 = \tfrac{35}{11}$$
$$y_2 = \tfrac{18}{11}$$
$$v = \tfrac{53}{11}.$$

**6** The dual problem in the variables $y_1$, $y_2$ is

Maximize $2y_1 + y_2$

subject to $\quad y_1 + 2y_2 \leqq 12$

$$4y_1 + 3y_2 \leqq 24$$
$$3y_1 + \phantom{3}y_2 \leqq 15$$
$$y_1, y_2 \geqq 0.$$

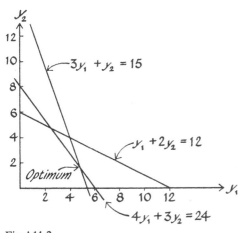

Fig A11.2.

The solution to this problem can be obtained graphically as shown in Fig. A11.2.

$y_1 = \tfrac{21}{5}$, $y_2 = \tfrac{12}{5}$ giving a maximum value $\tfrac{54}{5}$.

Also it is clear that the constraint $y_1 + 2y_2 \leqq 12$ is not binding. Therefore if we introduced slack variables $y_1^*$, $y_2^*$, $y_3^*$ into the three equations, $y_1^*$ would be non-zero in the solution whereas $y_2^*$, $y_3^*$ would have negative relative cost coefficients. This implies that $x_2$, $x_3$ are in the basis of the primal solution and hence we can solve the pair of equations

$$4x_2 + 3x_3 = 2$$
$$3x_2 + \phantom{3}x_3 = 1$$

to obtain $x_2 = \tfrac{1}{5}$, $x_3 = \tfrac{2}{5}$ as the optimum solution to the primal with an objective function value $\tfrac{54}{5}$.

**7** Assuming that both constraints are binding, we will have the primal-dual relationship at the optimum

$$z = c_1\bar{x}_1 + c_2\bar{x}_2 = b_1\bar{y}_1 + b_2\bar{y}_2$$

where $z$ is being minimized. The dual variable values $\bar{y}_1$, $\bar{y}_2$ measure the net increase in total cost of ingredients if constituents 1 or 2 were to be increased by 1 unit. The dual variables (or shadow prices) therefore imply a marginal value or cost of maintaining the constituent levels. Furthermore, in a larger problem we could assess whether the ingredients are correctly priced. Suppose we increased $x_1$ by 1 unit, this would imply increasing $b_1$ by $a_{11}$ and $b_2$ by $a_{21}$ in the equations, and the implied value would be $a_{11}\bar{y}_1 + a_{21}\bar{y}_2$. We could now compare this with the market price $c_1$ of $x_1$ to obtain a criterion for deciding which ingredients can be used at a profit.

**8** First it should be noted that

$$x_1 x_2 = \tfrac{1}{4}((x_1 + x_2)^2 - (x_1 - x_2)^2),$$

therefore by introducing the two new variables $y_1$ and $y_2$ together with the constraints

$$y_1 - x_1 - x_2 = 0$$

$$y_2 - x_1 + x_2 = 0$$

the cross-product term can be represented as a separable function and the extra constraints are also in the separable form.

**9** If the maximum value is $v$ we determine the least power of 2, say $2^m$ such that

$$2^m \geqq v.$$

Then the variable $x$ constrained to lie between 0 and $v$ can be expressed in terms of $(m)$ zero-one variables $y_i$ as

$$x = \sum_{i=0}^{m-1} 2^i y_i.$$

For example,

$$19 = 1 \times 2^0 + 1 \times 2^1 + 0 \times 2^2 + 0 \times 2^3 + 1 \times 2^4$$

$$= 1 + 2 + 16$$

$$= 19.$$

# Index

**A**

Allocation
  problems, 153
  of a single resource, 108
  of two resources, 116
Approximation programming, 91
Auxiliary variables, 164

**B**

Basis, 45, 49
Branch and bound methods, 26, 123
Branching policy, 126

**C**

Calculus, 29
Canonical form, 47, 51
Cargo loading problem, 111
Constraints, 19
  discrete, 21
  equality, 19
  inequality, 20
Created response surface technique, 80

**D**

Degeneracy, 61
Descent methods, 72
Differentiation, 6
Direct search procedures, 84

Disjoint alternative regions, 22
Dual problem interpretation, 172
Duality, 168
Dynamic programming, 26, 101
  backwards calculation of, 105
  method: forwards calculation, 103

**F**

Formulation, 161

**H**

Heuristic techniques, 26, 152

**I**

Initial solution, 54, 67
Integer programming, 134

**K**

Knapsack problem, 131

**L**

Lagrange multipliers, 35, 39
Linear functions, 42
Linear programming, 26, 42
  fundamental theorem of, 45, 61
  standard form of, 43
Local maxima and minima, 37

222    *Index*

**M**

Matrices, 11
Multi-stage decision processes, 102

**N**

Network, 13, 106
Non-linear optimization, 26, 65

**O**

Objective function, 18
Optima
  global, 23, 66, 143
  local, 23, 66, 143

**P**

Permutations, 12
Permutation
  exchanges, 146
  problems, 141
  procedures, 26, 141
Projected gradient method, 75
Projection matrix, 96

**R**

Recurrence relations, 104
Region, 10
Reliability problems, 113
Response surface technique, 97
Revised simplex method, 58

**S**

Scheduling problem, 127, 147
Separable programming, 88
Simplex method, 45
Simplex tableau, 56
Steepest descent, 69, 95
Stochastic problem, 119
Suboptimization, 138, 141

**T**

Taylor series, 8
Timetabling, 158
Transformation, 51, 166
Travelling salesman problem, 156
Tree searches, 123